EARLY CHILDHOOD
MATH ROUTINES
EMPOWERING
YOUNG MINDS TO THINK

EARLY CHILDHOOD
MATH ROUTINES
EMPOWERING
YOUNG MINDS TO THINK

Antonia Cameron

WITH PATRICIA GALLAHUE AND DANIELLE IACOVIELLO

Stenhouse
PUBLISHERS

www.stenhouse.com

Portsmouth, New Hampshire

Image in Figure 6.7 by Elli Stattaus from Pixabay.

Painting used in Figure 1.3 *Drina House in the Morning Mist* was created by Eliza Donovan. www.elizadonovanart.com.

Library of Congress Cataloging-in-Publication Data:

Names: Cameron, Antonia, author.
Title: Early childhood math routines : empowering young minds to think / Antonia Cameron with Patricia Gallahue and Danielle Iacoviello.
Identifiers: LCCN 2019036356 (print) | LCCN 2019036357 (ebook) | ISBN 9781625311832 (paperback) | ISBN 9781625311849 (ebook)
Subjects: LCSH: Mathematics–Study and teaching (Early childhood)
Classification: LCC QA135.6 .C359 2020 (print) | LCC QA135.6 (ebook) | DDC 372.7/044–dc23
LC record available at https://lccn.loc.gov/2019036356
LC ebook record available at https://lccn.loc.gov/2019036357

Cover design, interior design, and typesetting by Page2, LLC, Wayne, NJ.

Manufactured in the United States of America

PRINTED ON 30% PCW
RECYCLED PAPER

26   25   24   23   22   21   20   9 8 7 6 5 4 3 2 1

To my son, Seamus,

whose spectacular imagination and deep
wonderings sparked my curiosity about children's
minds and ignited my passion for teaching.

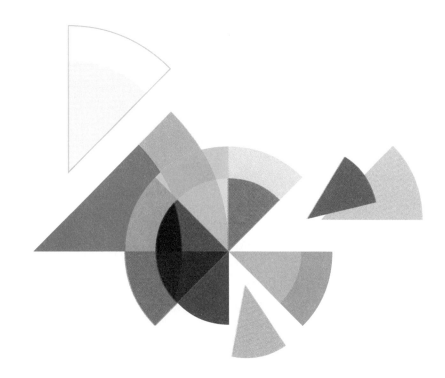

Contents

Foreword

Karen Economopoulos

This is a book that celebrates childhood and the power of young children's thinking. It is a book about children's ideas, the ones that make sense and the ones that don't quite make sense yet. This book focuses on the joyful learning that occurs when young children have opportunities to play with important ideas.

This is a book about purposeful teaching, because environments where children's thinking is at the center of the work don't just happen by accident. It is a book about teachers stepping back and children stepping forward.

This is a book about giving agency to children because when young students are engaged in interesting and meaningful work, they own the problem, they own the thinking, and they own the solution. This is a book about early childhood math routines that empower young students to think.

For those of us who have spent time studying the social, emotional, and cognitive development of young children and have been immersed in classrooms that foster this growth, the vignettes and images that Toni Cameron and her colleagues present will be familiar and affirming. For those of us who have spent time studying the complexities of early childhood mathematics and thinking about how to engage young students in making sense of the big ideas in early number, Cameron's notion of using routines to connect children to mathematics will be exciting and powerful. For others of us who might not have ever experienced how young children become counters and comparers and problem-posers and problem-solvers, this book will convince us that young children can and do think like mathematicians.

This is a book that teachers, coaches, and administrators can use tomorrow because it is about routines that are familiar and accessible. Many of these routines build on existing practices that are part of early childhood classrooms, such as taking attendance or counting the number of days in school. Toni, however, reshapes what is familiar and uncovers the potential mathematical opportunities often hiding in plain sight. She suggests variations that engage students and invite them into the routine. She opens up possibilities for engagement that shift the responsibility of the task from teacher to student. She offers questions that can reveal what students understand or do not yet understand. She comments on the math ideas and skills in accessible ways and in doing so, supports teachers to learn more about the math they are teaching.

This is also a book that teachers, coaches, and administrators can use over time to expand their practice, to improve their understanding of the big ideas in early childhood mathematics, and to hone their pedagogy as they observe students at work and listen to their thinking. Educators will connect with Toni's enthusiasm for what students say and do as they engage with routines described in this book. And it's hard not to get excited about Toni's commentary about why this work is important—important from a mathematical perspective, important from a pedagogical perspective, and important from an empowerment perspective.

This book gives teachers, coaches, and administrators a lot to think and talk about as they work together to improve their teaching of mathematics and empower our youngest mathematicians to think, learn, and do each and every day.

— Karen Economopoulos

Acknowledgments

This book is based on the accumulation of many experiences over my twenty-six years as an educator, and to acknowledge every single person who contributed to its creation would be a book in and of itself! That being said, there are a few people who have to be recognized because their contributions were of major significance, albeit each person was instrumental in different ways.

Let me begin by thanking Dani and Patricia, my collaborators, for their willingness to play and to try out the routines in their own classrooms. Their insightful feedback and ongoing support and friendship were critical at all stages of creation. It is also important that I acknowledge the support and encouragement my colleagues at Metamorphosis Teaching Learning Communities (MTLC) gave me in what seemed to be an unending journey of writing, rethinking, and revising. I thank Lucy West, my business partner and founder of MTLC, who often joked, "You're still writing the same book?!" Her humor encouraged me to keep on writing. I also thank my colleagues at MTLC who tried the reasoning routines in this book in various schools where they were coaching: Stephanie Slabic, Jenn Costanzo, Sonal Malpani, Ellen McCrum, Michael Cassaro, Renee McShane, and Deanna Catanzaro. They gave me invaluable feedback on what worked and what didn't, which then helped me rethink and refine some of the routines. My colleagues Alex Lawson, Anne Burgunder, and Betina Zolkower, professors from Lakehead University, Canada, New York University, and Brooklyn College, respectively, read and critiqued some of the chapters and offered specific readings to help me deepen my own knowledge of important research about mathematics teaching and learning. My friend, colleague, and writing partner from my days at Math in the City (The City University of New York). Sherrin Hersch, read and revised each chapter to make the writing more coherent. I am deeply indebted to my Stenhouse editors, Toby Gordon, who began this journey with me and retired, and Kassia Wedekind, who has been everything one dreams about in an editor. She has been discerning, thoughtful, appreciative, humble, and kind. She also has an astounding editing technique and can take the most ill-turned phrase and spin it into gold. I am also greatly appreciative of the Stenhouse team who supported me during the book's production stage: Amanda Bondi, my insightful production editor, and Shannon St. Peter, the production assistant who supported this book in myriad ways. I would also like to thank the other members of the Stenhouse team and say I am in awe of their talents (as evidenced in the stunning visuals of the book) and grateful for their collaboration and support.

Finally, this book would not have been possible without the support of my family. I especially thank my wonderful husband, James, who not only took over a lot of the household responsibilities in the last months of my writing but also knew when to be encouraging and when to keep silent, which often defused my writing doldrums. I also thank my ninety-two-year-old mother, Millie, who has been an ongoing source of inspiration. She always believed in my capacity to write, and she never gave up on me even when I was willing to give up on myself.

Companion Website Access Information

The *Early Childhood Math Routines* Companion Website includes access to reproducibles available for downloading and printing, slide decks for facilitating routines, and additional practice routines. To access the Companion Website please follow the directions below.

Access Information:

- Go to **earlymathroutines.stenhouse.com**
- Create your account. Follow the on-screen directions to create your personal account and enter this passcode: **MathRoutines002**

You will now have access to:

- Reproducibles (cards and gameboards) for use with the practice routines
- Over thirty slide decks to support facilitation of quick image and *Is It Fair?* routines
- Extras including additional practice routines, supplemental readings and a place value interview assessment

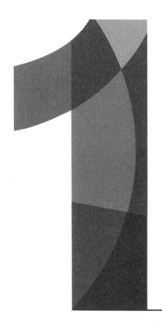

Transforming Early Childhood Math Routines

> " We don't stop playing because we grow old;
> we grow old because we stop playing. "
>
> **George Bernard Shaw**, author and playwright, *Pygmalion*, 1912

Wonder and the Early Childhood Classroom

If we believe that young children need time and space to wonder, create, and set goals for their own learning, we must structure our early childhood classrooms to reflect these beliefs (Duckworth 2006). The current reality is that many early childhood classrooms do not have *play* at the center, but rather are more focused on academic learning. This shift has happened in part because the current expectations for learning, which are set by state and local standards, have diminished the importance of play and have instead placed great emphasis on the development of specific skills (usually reading and math skills) in early childhood. Although some of the goals underlying these standards are admirable (e.g., to level the playing field so all children—no matter their socioeconomic background— learn to read), the real question might be "At what cost to children?"

The conundrum faced by educators of young children is this: How do we, given the current educational structures and expectations that are focused on meeting standards, find ways to create rich, vibrant learning communities where exploration and play are central? This is a question we (Toni, Dani, and Patricia) have grappled with in our own teaching and coaching. Although there is no easy answer to this question, one solution we have found is to use thinking routines in our teaching that are multilayered; they can be used to develop specific content ideas (and meet different standards) *and* they also can be seen as tools to develop playful mindsets, problem-solving habits of mind, and a wide range of communication skills.

Play and Learning: Ways of Seeing Are Ways of Being

Picture this: a birthday party; a child, after unwrapping presents, leaves all of them in a pile and goes off to play with the box they came in. All the bells and whistles, the bright colors, the expensive gizmos meant little to the child—they didn't entice her in the least. Instead she takes the plain brown cardboard box and it becomes the object of fascination; hours are spent engaging and reengaging with it. Why is this?

For most adults, to paraphrase Gertrude Stein's famous quip, a box is a box is a box (Stein 1922). A box is utilitarian; it is a container; you put things in it; you mail things in it—its potential is fixed. For children, on the other hand, a box is a gateway to another world where purpose and usefulness are directly connected to the imagination. A child wonders, "Could this object be a dollhouse?" She asks, "What if I put my boat in the stream? Will it float?" She builds a castle with the box, hides in it, or uses it as a car. The box's usage is limited only by the boundaries of what can be envisioned (see **Figure 1.1**).

The United States has a National Toy Hall of Fame.
The ordinary cardboard box has been inducted into the Hall of Fame and lives alongside other beloved toys like Lego and Mr. Potato Head!

Two key questions to consider before reading on:

1. Why is the give-a-present-and-the-child-plays-with-the-box-it-came-in experience so common?

2. What implications does this have for early childhood education?

FIGURE 1.1

A two-year-old makes a box her bed!

Because things are not fixed rigidly in children's minds, they can engage in the world and with the world in playful and imaginative ways. Their explorations are pathways to *knowing* and their learning is fueled by their questions and what-ifs. As children puzzle about things, as they test out their ideas and predictions, they come to understand the world around them.

Questions to Ponder

- What routines do you use in your classroom, and why do you use them?
- Are some routines more effective than others?
- What makes a routine effective or ineffective?

Routines as a Form of Play: A Tale of Two Classrooms

Classroom #1

Ms. Martin's kindergarten students gather in a circle in the meeting area for their morning routines. Ms. Martin also sits as a participant.

Ms. Martin: Good morning!

Children: Good morning!

Ms. Martin: Today we are going to start by doing our routine, *You Say, I Say*. [The children become visibly excited.] I know you really like this routine a lot—it's become one of our favorites. Who would like to start us off. Jacob? Okay, today we will go around our circle in a clockwise direction [demonstrates what this means with her hand]. This means after Jacob starts, Cali will go, then Maggie will go next. And remember the rules we came up with for this routine [points to a chart that lists the rules]. First, talk only when it's your turn. Second, each of us will say only one word. Third, the word we say needs to be connected to the word before. Can someone give us an example of this kind of connection?

Cali: The other day, when Tim started and said, "Duck," Milo said, "Quack." He said it like this, "Quack!" [imitates a duck honk; children laugh]

Milo: That was so funny!

Ms. Martin: Sometimes what we say is funny and we laugh. Sometimes what someone says is puzzling—it might make us think, "Hmm! I didn't think of that!" Sometimes what someone else says is exactly what we were thinking—and we can show our connection with our hand signal [demonstrates the hand signal for "I agree"]. Remember, this routine is about making associations—making connections. Our brains do that—they play with ideas; they play with words. Try, when you hear what is said, to say the first thing that comes into your mind. Let's try to let our words flow. Okay. Jacob, let's go.

Jacob: Blue.

Cali: You say blue; I say green.

Maggie: You say green; I say red.

Maya: You say red; I say barn.

Toi: You say barn; I say cow.

Axel: You say cow; I say milk.

Kai: You say milk; I say glass.

Gaia: You say glass; I say soda.

Ben: You say soda; I say burp.

[Children laugh loudly.]

Fran: You say burp; I say stinky.

Pete: You say stinky; I say feet.

Usha: You say feet; I say toes.

Ms. Martin's routine continues in this way until every child has spoken. As children share, she keeps track of their words in a notebook. When the routine is over, Ms. Martin will share the list of words on the board so the students can reconstruct and reflect on their word journey. (The final words after toes are *socks, shoes, dancing, party, cake, chocolate, candy, gum, chew, bubble, pop*.) After every child has said a word, Ms. Martin circles the words *blue, green*, and *red* and asks children what these words have in common ("They're colors"). Then she says, "This word [circles *barn*] puzzled me. After *red*, Maya said, *barn*. Turn and talk to your partner about where *barn* might have come from." After children talk, they share out.

Kato: We think barns can be red.

Gaia: And the other words go with barn.

Ms. Martin: What words?

Gaia: *Cow* and *milk*. We get milk from cows.

Ben: And *glass* goes with *milk*. You pour milk into a glass.

Don: And soda goes in a glass; when I drink it fast, I burp. [laughter]

Ms. Martin: Isn't it amazing? [points to the word list] We started with a color and ended with the word *pop*! What a surprising word journey we had today! Okay. Let's get ready to do our count around. Everyone stand up!

Classroom #2

It's 8:40 a.m. on a Tuesday morning in December and Ms. Perry's kindergarten class gathers on the rug for morning meeting. Ms. Perry says, "Good morning, kindergartners! It's time for our morning meeting. Let's start by looking at our calendar." A loud groan is heard from several children; other children giggle at the sound.

Ms. Perry continues, "Jamal, why don't you help us out today. You can use our special pointer." She hands him the pointer. "Remember to tell us what today is." Ms. Perry points to the days of the week that are written on laminated cards on the board. "Look at our calendar to see what the last day was that we were in school. Is today Monday? Tuesday? Wednesday? Thursday? Friday? Those are the days we are in school. After you tell us what today is, then tell us what day yesterday was and what day tomorrow will be."

Jamal points to the calendar and says, "Today is Tuesday. Yesterday was Monday and tomorrow will be Wednesday."

Later in the morning at a grade-level meeting, Ms. Perry expresses her frustration with this routine and wonders aloud to her teacher colleagues, "Why are my kids so bored with *The Calendar* routine? I keep thinking, 'What am I doing wrong? How can I fix this routine so kids like it?'" No one offers any suggestions, and Ms. Perry continues with this routine because it is one of the expectations outlined in the curriculum adopted by her school and district.

Two classrooms, two different routines, and two different learning experiences for children. What happened to create such radically different learning outcomes? Let's begin by stating that this comparison isn't about the teachers being "good" or "bad." Both teachers are thoughtful, well meaning, and hardworking. They both want their students to engage in meaningful mathematical experiences. However, something was amiss in Ms. Perry's classroom. She recognized it and asked her colleagues how she might improve the routine. She wondered how to make this routine more engaging for her students. And although "fixing" a routine to make it more engaging to children is an admirable teaching behavior, maybe the real question isn't how to make a routine better, but *why* do this routine in the first place and what exactly are children learning? To explore this question, let's compare the two routines, *You Say, I Say* and *The Calendar*. Let's think about the similarities and differences between the routines and how their differences might impact children's emotions and learning (**Figure 1.2**).

FIGURE 1.2 Comparing Routines: *You Say, I Say* and *The Calendar*

Similarities	Routine	Example
Both have clear repetitive structures that children internalize to make the flow of the routine effortless. Repetition is used to help children internalize the routine's structure.	1. *You Say, I Say*	You say [repeat previous word said]; I say [add the word that comes to mind].
	2. *The Calendar*	Today is [day of week]. Yesterday was [day of the week]. Tomorrow will be [day of the week].
Clear expectations are set for children's behaviors.	1. *You Say, I Say*	Children know they will speak one after another and that their job is to listen to what is said and say the first word that comes to mind. There is no calling out or cross-conversation.
	2. *The Calendar*	Most children sit in the meeting area, and a few students go through the routine, naming the appropriate days that answer the three questions. There is no calling out or cross-conversation.

Differences	Routine	Example
There is a big difference in the type of intellectual engagement required of students. In *You Say, I Say*, the structure is a canvas on which students paint with their imaginations; in *The Calendar*, there is only a canvas and no paint.	1. *You Say, I Say*	Children use the routine structure to make word associations. They can use classification as a structure (e.g., using color words) or they can use a different *kind of association* (e.g., a barn is red). The variation in what is said is exciting and humorous. This kind of imaginative play creates a different kind of energy in the classroom—one in which children are attending to what's being said by their peers.
	2. *The Calendar*	The rote recitation (the memorization of the days of the weeks, months in a year, etc.) does not engage the children in intellectual play. There is a very limited repertoire of responses (e.g., the seven days of the week) and this kind of learning quickly becomes dull once a child knows the days of the week—there's nothing to learn, just repeat.
The teaching strategies vary greatly. In *The Calendar*, social knowledge is shared. In *You Say, I Say*, the teacher revisits what's been said to help children think about how the words are connected. In doing this, she is developing and expanding their reasoning strategies.	1. *You Say, I Say*	At the end of the routine, the teacher writes the words children said on the board. She circles several words and asks students what they have in common (e.g., they're colors). She also uses a discourse move (turn and talk) to slow down the processing and give all the children a chance to share their ideas with a peer.
	2. *The Calendar*	The routine is perfunctory. It is structured in a way that leaves little opportunity for student reasoning or talk.

We recognize that there are many ways to use the calendar to engage students and that the snippet of conversation from Ms. Perry's class focuses on a rote learning experience. For example, we have seen teachers make the routine more engaging by putting children's birthdays and special events on the calendar and posing questions about these special days (e.g., "How many days until . . . ?"). We have also seen teachers add patterns to the calendar pieces (e.g., square, triangle, circle, square, triangle, circle) as a way to enliven the routine. Whatever you do with your calendar routine, here are several questions to consider: How engaged are your students when you do this routine? What ownership do they have over the routine? What exactly are your students learning? Might the time devoted to *The Calendar* routine be used more productively? For more on why teachers might rethink *The Calendar* routines, see **Appendix A**: A Math Note on *The Calendar*.

Teacher Note

It's Not Just the Routine: Teaching Matters Too!

It is not *just* the quality of a routine that matters. A wonderful routine is only as effective as the teaching that goes along with it! This means the quality of a routine is directly connected to what a teacher believes about learning (what she values) and the teaching tools (the pedagogy) she uses to bring her vision to life. If a teacher believes children learn best when they are given problematic situations to puzzle about and are asked to share their ideas publicly, she needs specific pedagogical tools (e.g., like how to facilitate talk) to make this vision a reality (National Council of Teachers of Mathematics 2014). Throughout this book we will explore powerful pedagogical moves that help us get the most out of our math routines with young children.

Routines in the Early Childhood Classroom: How We Define Routines

If you look up *routine* in a dictionary, some of the nouns that come up are *procedure*, *pattern*, *drill*, *system*, and *method*. Some of the adjectives connected with routine are *commonplace*, *conventional*, and *ordinary*. Although these definitions hint at what might make routines boring, classroom routines do not have to be uninspired.

We define routines as a kind of *system* that contains a predictable *pattern* that makes them easy for students to learn and remember. Repetition is an important feature of routines; it helps students internalize the routine's structures (Ritchhart et al. 2006). As students become more familiar with the routine, the structure becomes automatic. For example, the structure of the *You Say, I Say* routine will stay the same over time (although at some point, children may not need the sentence frame of "You say . . . , I say . . . "). However, because the words children say change every time they do this routine, the thinking never becomes rote. Although the structure (writing the words generated on the board and looking for connections) stays the same, the actual connections children make are highly variable.

Even though different routines may have similar structures, the purpose of one routine may be radically different from another. For this reason, it is important for teachers to be able to distinguish between types of routines and also to be clear about their purpose as they plan routines.

ROUTINE TYPES

Educational researchers Leinhardt, Weidman, and Hammond (1987) have organized routines into four broad categories:

- **housekeeping routines** (e.g., how to put away materials in a classroom);
- **behavior management routines** (e.g., how children walk to the rug and gather in the morning meeting circle);
- **discourse routines** (e.g., norms for how members of the classroom community listen to each other and ask questions of one another when they don't understand something that has been said);
- **learning routines** (e.g., a math routine in which students figure out how many children are present and how many milks they need for snack).

Although it's helpful to be familiar with these types of routines, it is also important to recognize that these are not rigid classifications; a single routine may fit into more than one category. All of the math routines in this book are structured to develop multiple social and learning behaviors. For example, here are a few things children learn as they participate in counting routines: (1) physical behavior: how to sit together and interact socially in the meeting area; (2) discourse: how to listen to each other—attend to the numbers other children are saying in the sequence—and let others know if they agree or disagree with what's been said; (3) math learning: the rote counting sequence and patterns of counting numbers in base ten.

Routines as Mirrors of Our Values

It is no secret that carefully thought-out and well-managed routines are essential for the smooth running of classroom life. Moreover, the routines a teacher selects communicate what is valued in the classroom. If students are given many opportunities to talk to one another during turn and talks and collaborative tasks, they learn that their ideas are worthy of consideration and that they can figure things out by talking to each other. If, on the other hand, student talk is focused primarily on answering only the teacher's questions, students may think that their job is to figure out what the teacher is thinking or what she wants to hear.

Although classroom routines are essential for creating a structured environment where students can thrive, not all routines have positive long-term effects. It is important for us, as teachers, to think carefully about the explicit and implicit messages that underlie our routines. Here's an example of the dual nature of routines: one common classroom routine is for students to raise their hands and wait to be called on by the teacher before speaking. This routine may, at the outset, seem positive—it creates order by structuring the learning environment in ways that help students manage their impulsivity. But there can also be negative consequences to hand-raising routines. Always relying on hand-raising can encourage compliance above all else. Even though compliance may not be the intended outcome, children may internalize a classroom hierarchy that places the ultimate source of power in the teacher. For example, in whole-group discussions when children have internalized the "I-can't-speak-until-I've-gotten-permission-from-my-teacher" mentality, there is often a stilted kind of conversation. Whether they agree, disagree, or have a question, children are not able to speak directly to the person expressing an idea. A child's personal power can also be diminished when he realizes that he needs permission to speak; that act of control on the teacher's part—even when well intentioned—can have negative consequences on the minds of learners (Toshalis and Nakkula 2012). One way to avoid developing this kind of negative behavior is if we, as teachers, are careful to balance hand-raising structures with ones that encourage more natural conversational structures. The latter provide opportunities for student choice and are essential in developing children's autonomy.

ENVISIONING YOUR ROUTINES

As you think about which routines you use and why you use them, try to picture your ideal classroom: What does it look like? Sound like? Where are you in the classroom and how do students relate to you and each other? Once you have a clear vision of what you want, you can think about your choice of routine by posing the questions: How will this routine help you create a classroom culture that is a true reflection of what you value as a teacher? What evidence are you looking for in children's learning and behavior that will help you assess its effectiveness?

CAREFULLY CHOOSING YOUR ROUTINES

As we saw in the comparison of *I Say, You Say* and *The Calendar*, not all routines are created equal. Some routines create rich talk and deep thinking; others do not (Ladson-Billings 2014). Some engage students in playful thinking; others make them disinterested. How do we then, as teachers, choose routines or decide what constitutes an effective routine? Here are a few questions that might help with analysis and selection. These questions focus on the *what*, *why*, *who*, and *how* and are essential in lesson design (West and Cameron 2013).

1. Why are you using this particular routine—what is its purpose?
2. If the routine is designed to develop mathematical thinking, what are the big ideas or strategies underlying or embedded in it?
3. Who are your students? What are their developmental needs? Is this routine appropriate for all students? For some students? Which ones?
4. What are the possible learning pathways (Clements and Sarama 2014) for students in this routine? Is it differentiated in ways so that all learners will have access?
5. How will this routine meet or challenge your students' developmental needs?
6. How do you assess and document student learning?

ROUTINES SHOULD EVOLVE OVER TIME

It is helpful to keep in mind that all routines can (and perhaps should!) evolve over time. As children engage in a routine multiple times, they internalize its structure (Ritchhart et al. 2006). Over time, some of the teacher scaffolding becomes unnecessary. It is important to be aware of opportune moments when you can shift your routines—this prevents them from becoming mechanized and boring. For example, with *You Say, I Say*, teachers have noticed that over time, their students do not need to use this sentence frame, because they have internalized the routine's structure and can easily share their words without it and still maintain the flow of language.

Learning Pathways

Learning pathways can be thought of as the possible directions that students can take in their mathematical journeys within a given routine. It is important for teachers to consider these different learning paths and to be prepared to slow down or extend a routine based on student needs. This type of planning is possible only when we deeply understand the content embedded in a task. To help us think about children's learning journeys in this book, we will refer to a developmental framework called a *landscape of learning*, which is made up of big ideas, strategies, and models. This model was developed by Maarten Dolk and Cathy Fosnot. To read more about this tool and how it might be used, see, *Young Mathematicians at Work: Constructing Number Sense, Addition and Subtraction* (Fosnot and Dolk 2001).

THE ROLE OF THE TEACHER IN SETTING UP ROUTINES

Teacher intent and clarity about purpose (Why am I using this routine?) and ability to notice (Pam and Amari distract each other and may have to be moved to different spots in the meeting area) are foundational for routines to run smoothly. Equally critical to the successful implementation of routines are the choices the classroom teacher makes about when to slow down and what to highlight (e.g., In a counting routine, why it's helpful to sit a certain way; how to listen carefully so that you know what number to share when it's your turn; how to disagree with a mistake in the count without being mean). The more clearly the teacher envisions *how* the routine works, the more easily she can tweak specific aspects of the routine when they are not working and find ways to help students develop as individual learners and as members of a community (Barell 1991; Hiebert et al. 2000).

What Kind of Routines Can You Expect in This Book?

This book, in focusing on routines, emphasizes an important subset of discourse and learning routines called thinking routines. "Thinking routines are easy-to-learn structures, mostly taking the form of simple sets of questions or metaphors that naturally involve children in *thinking processes*. These routines may expand the children's repertoire of cognitive strategies because they constitute a major form of organizing memory and thinking" (Salmon 2011, 366).

The work in this book builds on the spirit of thinking routines first described by Harvard's Project Zero. According to the work of this project, thinking routines "focus on the establishment of structures that weave thinking into the fabric of the classroom and help to make the thinking of everyone in the classroom more visible and apparent. These kinds of routines provide models *of* thinking and are often a vehicle for incorporating thinking language into classrooms as well" (Ritchhart et al. 2006).

Thinking Routines in the Early Childhood Classroom: *Notice and Wonder*

The beauty of the thinking routines is twofold. First, they are quick and enjoyable for children to learn and do. Second, they hold great potential for developing students' mathematical reasoning and communication skills. Here's an example of a thinking routine created by the educators and researchers at Project Zero that we have adapted and used in our classrooms. They call it *See, Think, Wonder*. We call it *Notice* and *Wonder* (Ray-Riek 2013).

Teacher Note

A Look at Research

Vygotsky is the preeminent proponent of the position that pretend play increases intelligence and develops abstract reasoning. Vygotsky uses his famous example of the stick that, in play, becomes the horse, to explain how play allows children to develop a separation between perception and meaning. The stick is the "pivot," which allows thought, word meaning, to be separated from objects and action to arise from ideas as opposed to arising from things. Although the stick is still needed to separate thought and object, the child's relation to reality is now changed because the structure of his perceptions has changed. For the first time, meaning predominates over object (Vygotsky 1978). Vygotsky writes that "this characterizes the transitional nature of play; it's a stage between the purely situational constraints of early childhood and adult thought, which can be totally free from real situations" (1978, 98).

NOTICE AND WONDER: DEVELOPING PLAYFUL MINDSETS

The beauty of the *Notice and Wonder* routine is that it can be used to help students generate thinking by focusing on what they notice in a picture, followed by sharing their wonderings with others. *Notice and Wonder* has the potential to do many things simultaneously. It can be used to develop children's language, reasoning, and communication skills. It can also be used to develop two important problem-solving habits: slowing down to notice and then using what is noticed to generate questions. In this routine, children's questions and wonderings are a form of intellectual play and can be seen as a way to support their creativity. Let's use the **Table 1.1** dialogue box on page 15 to examine the structure of the routine and think about how it develops children's language and communication skills.

After gathering her students on the rug, Ms. Casey explains that the class will learn about a new routine called *Notice and Wonder*. After exploring what the class knows about the words *notice* and *wonder*, Ms. Casey reveals the painting, *Drina House in the Morning Mist* (see **Figure 1.3**), and asks the students to share what they notice. Several children respond, "I see a house," "I see trees," and "I see water." But then there is a pause. Ms. Casey nudges more children to share their ideas, reinforcing, "There are no wrong answers. Everyone's noticing is important." The children begin to expand upon their noticings. "The house is on a rock in the water!!" "I see yellow trees; green ones too." Let's join the class as they begin the wonder portion of the routine. To read this classroom dialogue in full, see **Appendix D**.

FIGURE 1.3   *Drina House in the Morning Mist* **by Eliza Donovan**

TABLE 1.1 *Notice and Wonder*: Analyzing How a Teacher Uses a Routine to Facilitate Student Reasoning

	Transcript of *Notice and Wonder* Routine	Analysis of Teaching	Implications for Learning
1	**Ms. Casey:** [As children have shared their noticings, she has written them down on the board.] Thank you all for sharing your noticings. I'm going to read what I wrote on the board; read with me if you want [reads the list of words and phrases]. So many noticings! Now we're just going to think about our wonderings. What do you wonder about all the things that we've noticed? [Hands shoot up right away.] Let's give ourselves some time to think. Please put your hands down, because we all need time to think about our questions, our wonderings. [Hands go down; she gives students a few minutes to generate questions.] Are we ready? Who would like to start off our wonderings list? Maya?	The teacher uses a talk move, wait time, to ensure that all learners have time to create questions. This simple move provides access to all learners. It also helps children realize that thinking isn't about *speed* (the person who speaks first isn't viewed as the smartest or the best). Slowing down to create thinking space is a necessary step in developing a culture of ideas.	Understanding takes time. As students realize that they can think about their thinking and refine their ideas and questions by framing them first in their own minds, they begin to develop the kind of self-control necessary for prolonged, deep thought. Think of this habit of mind as a critical tool for developing effective problem-solving strategies.
2	**Maya:** Are we doing the popcorn again?	As children internalize the routine, they are able to name the expectations. Notice how the word *popcorn*, which at first was met with laughter, is now understood as a metaphor for a way of speaking.	
3	**Ms. Casey:** Yes. Thanks for reminding us. We'll popcorn out our wonderings. Please wait until I finish writing before you speak, though, because I had a hard time keeping up with everything you noticed.	Notice how the norms are not all set at the beginning as a list, but emerge as the need for them arises. The teacher sets a new norm for *when* to popcorn out—wait until I've finished writing or else I can't keep up. She also gives students a rationale for what she's asking them to do, which signals that it's important for them to understand her choices—they're not arbitrary.	
4	**Maya:** Why is the house in the water?		
5	**Child:** Who built the house?		
6	**Child:** How could someone build a house in the water?		
7	**Child:** Why would someone want to live there?		

(continues)

TABLE 1.1

(continued)

	Transcript of *Notice and Wonder* Routine	Analysis of Teaching	Implications for Learning
8	**Child:** Do they dive off the porch into the water?		
9	**Child:** Do they fish from the porch?		
10	**Child:** How did they make stairs in a rock? That must have been hard to do.		
11	**Child:** Do they know how to swim or do they have to wear that round thing?		
12	**Child:** Where do they get their food?		
13	**Child:** How many people live there? It's so teeny.		
14	**Child:** Will the house go in the water if there's a big storm?		
15	**Child:** How do they stay warm in the winter?		
16	**Child:** Are they sad by themselves on the lake? Where are their friends?		
17	**Child:** If the water goes up, will it go into the house?		
18	**Child:** Are they scared to be out there by themselves? What if there's lightning?		
19	**Ms. Casey:** Any other wonderings? Okay. This is our first time doing a *Notice and Wonder* routine, but we'll be doing this again. How many of you enjoyed this new routine? [Children give a thumbs up signal.] Here's something we can think about as we continue to use this routine: how do we use what we're noticing to create our questions, to have wonderings? To notice you have to slow down and really observe and think about what you're seeing. Look at all the things we noticed. [She circles the words that describe the house.] Look at all the ways we described the house: It has windows. It's red. It has a porch. It has two windows and a green roof. Now look at some of our questions about the house: How did someone build a house on the water? And why do they want to live there? Notice too that we wondered about the safety of the house. Would it fall into the water, and would people be warm? Would they get lonely? What wonderful questions we raised about a picture we might not have even noticed if we hadn't slowed down to pay attention and wonder!	The teacher reframes the purpose of the routine. She revisits the noticings and highlights the words used to describe one thing: a house. As she does this, she emphasizes complex language structures (e.g., it's a red house with windows and a porch). She also highlights the importance of slowing down to notice *before* generating questions. The more we notice, the more varied our questions will be. The emphasis on the importance and power of questions is a critical goal for learning.	Over time and with repetition, children's descriptive language improves with the *Notice and Wonder* routine. As they think about what they notice, they begin to use adjectives to give their observations more detail (e.g., "a tiny red house" rather than just "a house"). Notice too that their questions run the gamut from just three noticings to many. These questions could be explored by children in a small- or whole-group writing exercise. A question like "Who lives in the house?" could generate a story . . . Once upon a time, on a house in a lake, lived three children who loved to swim . . .

Why Notice and Wonder?

Noticing and wondering are at the heart of learning and the heart of mathematics. As a routine, *Notice and Wonder* is a form of imaginative play. With this routine, very powerful learning behaviors can be developed. By structuring the activity initially to just focus on noticing, the ideas that arise can become more nuanced and detailed as children slow down to observe for themselves and then listen to each other's noticings (Csikszentmihalyi 1991). After noticing, children may generate questions (their wonderings), which might be concrete, whimsical, or ingenious (Donaldson 1986). Although the immediate outcome of this type of routine is the development of children's oral language and communication skills (by describing what's seen and listening to the noticings of others), there are broader implications for children's learning as well. *Notice and Wonder* develops habits of mind that are essential to all learning and are specifically critical in mathematics and problem solving. Cuoco, Goldenberg, and Mark (1996) point to specific habits of mind that are important to mathematics. They encourage teachers to provide opportunities for students to be

- **problem sniffers** (take delight in finding hidden patterns);
- **experimenters** (play with problems they encounter);
- **describers** (give precise descriptions);
- **tinkerers** (develop the habit of taking ideas apart and putting them back together);
- **visualizers** (visualizing relationships, processes, change); and
- **conjecturers** (making conjectures on the basis of data).

In mathematics, successful problem solvers utilize these habits of mind or learning dispositions. When students slow down to make sense of a problem, they begin to notice, pull things apart, examine details, and determine what information is and isn't relevant. All of these actions comprise noticing. As students notice, they also develop another important problem-solving habit of mind, the ability to pose questions, first about what one is noticing and then, as one engages in the problem, about what one is thinking and learning. This unpacking of a problem is directly connected to playing with it, the ability to imagine, refine, wonder, and rewonder.

As you read this book, you'll learn routines to develop big mathematical ideas in young learners. Although this book focuses on routines to develop math thinking, the practice of noticing and wondering is encouraged throughout all of the routines. Our goal is for students to develop mathematical understanding while also developing general curiosity about the world around them. You will also find helpful teaching tips that are designed to support you as you play with these routines in your own classroom. And perhaps most importantly, you will learn how to build a community of problem-solvers who view math as beautifully complex and creative, and you will notice this focus on collaborative thinking and reasoning infused in the math routines throughout this book. In Chapters 2, 3, and 4, we share routines that help children build big ideas in early number and place value. Specific mathematical models (e.g., the ten-frame, the bead string, the growing number line, and the hundreds chart) are used to support and deepen students' reasoning. In Chapter 5, we use the routine *Is It Fair?* to develop children's reasoning in comparative situations (e.g., *I have six pretzels; Ann has eight. Is it fair? If it's not, how could we make it fair?*). This kind of reasoning is the foundation of developing part-whole relations, which is an essential component in developing fluency in addition and subtraction. In Chapter 6, we examine how visuals called quick images can be used to develop children's ability to describe what they see (communicate how an image is organized in space and specific kinds of groupings) and develop more efficient strategies (move their number strategies beyond counting by ones to reasoning with key relationships like doubles, near doubles, and "friends of ten"; i.e., the multiples of ten in "kid speak").

We wish you a playful learning journey as you explore the routines in this book and bring them to life in your classroom with your students.

2 Re-envisioning *How Many Days Have We Been in School?*

" If you always do what you've always done, you'll always get what you've always got. "

Source Unknown

Many of us are familiar with whole-group routines in which children keep track of the days of school (Shumway 2011). Like many other calendar routines, keeping track of the days of school often becomes rote and uninteresting, for teachers and students alike (Stipek 2018). Our *How Many Days Have We Been in School?* routine could not be more different. (See **Figure 2.1** for one teacher's set-up for this routine.) In this routine, students are invested leaders as they engage in and access different representations of the number of days they have been in school.

FIGURE 2.1 One teacher's set-up for the *How Many Days Have We Been in School?* routine

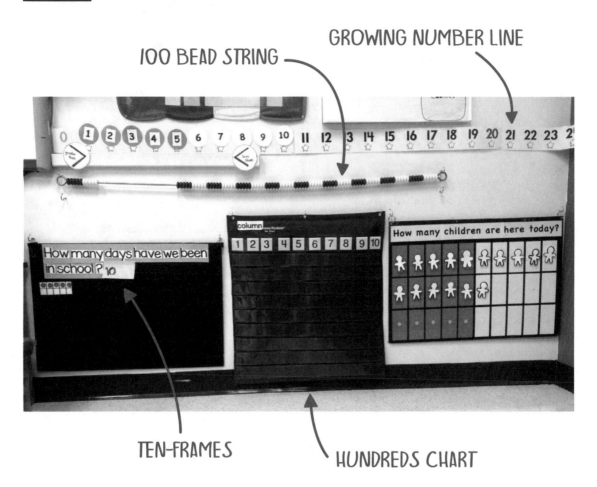

In contrast to other routines that involve counting school days (e.g., ones that use bundling straws or other objects to represent place value), the power of this routine is directly related to the models it utilizes—the bead string organized in fives, the ten-frame, the growing number line, and the hundreds chart. These models support students as they develop an understanding of our base-ten number system. The routine's structure, the fact that there is a real reason to keep track of how many days we've been in school ("We don't want to miss the one-hundredth day and our celebration"), and the use of different mathematical models all support students' reasoning with numbers (Gravemeijer 1999; Klein 1998). This use of different models also helps to differentiate learning. Although some children make sense of the "How many days of school?" question by using the growing number line, others gravitate toward using the ten-frame model to make sense of the context. Over time these models cohere into a representational system as children start to recognize that a number can be shown in different ways (e.g., thirty-nine can be a number on a number line; it can be represented as three filled ten-frames and another ten-frame with nine dots).

In this routine, each model is introduced separately. This introduction over time maximizes students' experiences with each model and is a way to help them build connections between the models. Following is the basic structure of the routine.

DAY 1: INTRODUCING THE BEAD-STRING MODEL AND THE CONTEXT FOR THE ROUTINE

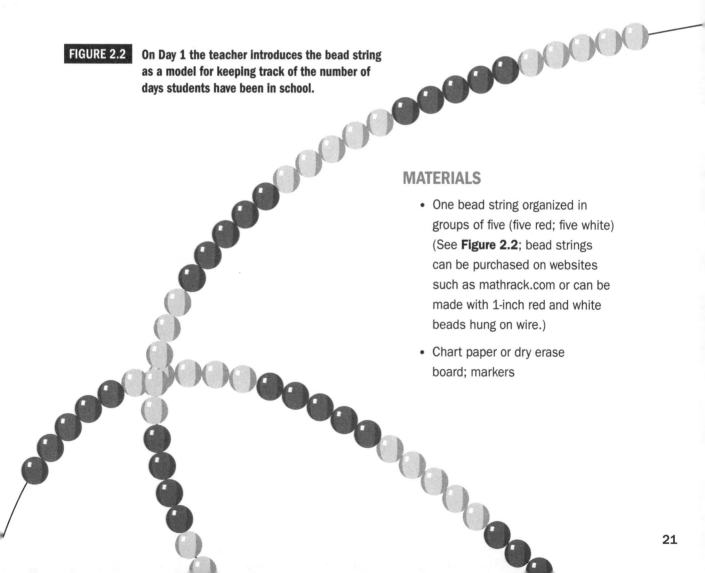

FIGURE 2.2 On Day 1 the teacher introduces the bead string as a model for keeping track of the number of days students have been in school.

MATERIALS

- One bead string organized in groups of five (five red; five white) (See **Figure 2.2**; bead strings can be purchased on websites such as mathrack.com or can be made with 1-inch red and white beads hung on wire.)

- Chart paper or dry erase board; markers

🎯 GOALS FOR DAY 1

The teacher introduces the bead string as a model and shows children how to use it. Students explore the structure of the model. The teacher can establish these goals by saying to the students, "This bead string will be a tool to help us keep track of how many days we have been in school. The reason we want to know how many days we've been in school is that on the one-hundredth day of school we will have a big celebration!" The power of this context—and all real contexts for students—cannot be underestimated in the development of powerful mathematical thinking (Van den Heuvel-Panhuizen 2001).

INTRODUCING THE MODEL AND EXPLORING THE STRUCTURE

1. Ask students, "What do you notice about this bead string?" Possible responses will either focus on the visual (how it looks) or the quantity (how many beads are on the bead string). Students might say,

   - "There are red and white [things, beads, balls, circles]."
   - "There are a lot of [things, beads, balls, circles]."
   - "It looks like a necklace."
   - "There's a pattern: red, white, red, white."
   - "It hangs."
   - "It's on wire [string]."
   - "The beads move."
   - "There are five red and five white beads."
   - "It goes in ten."
   - "There are one hundred beads."

2. Because you do not want this to take a lot of time, chart children's responses and refrain from expanding upon their comments.

3. Then tell students, "We are going to use the bead string to help us keep track of how many days we've been in school. We are going to move one bead each day we are in school. We will move the beads from right to left [demonstrate what this looks like]. This is the left side of the bead string. This is the right side of the bead string. I move a bead from right to left. We are keeping track of our days in school because on the one-hundredth day of school we are going to celebrate and have a big party!"

Teacher Note

Bead String Facilitation Tip

Even if a child says, "There are one hundred beads," refrain from asking, "How did you know?" on Day 1 of the routine. The reason for *not* exploring the quantity at this time (even though it's impressive that a young child might be able to count one hundred beads!) is that the goal of the introduction is to focus briefly on *noticings* (Ray-Riek 2013). Also, if you do decide to explore the "How many?" question, you might find that students become restless when the explanation takes too long or is too far beyond their understanding. Since this is the first day of school, some children will need to build up the stamina to sit for prolonged periods of time. One way to manage the one hundred beads comment is to write it down on chart paper as a conjecture or an idea about what *might* be true (Mason, Burton, and Stacey 2010). "Mike thinks there are one hundred beads. That's an interesting idea, and I'll write it down here for us to all think about later."

See **Figure 2.3** for an example of another conjecture developed by kindergartner Lucia through *The Calendar* routine. In this case, Lucia's teacher made a regular habit of listening for students' hypotheses around big math ideas (especially connections students made about the structure of the models and the "How many days have we been in school?" question), writing down and posting these conjectures publicly (Simon 1995).

FIGURE 2.3 Lucia makes a conjecture during *The Calendar* routine that connects the structure of the bead-string model with the ten-frame model.

Lucia's Quote

"We already used the beadstring to prove how many days we have been in school. We can count the same way on the 10 frames."

DAY 2 TO DAY 5: USING THE BEAD STRING TO KEEP TRACK OF THE DAYS IN SCHOOL

MATERIALS

- Bead string to one hundred
- Chart paper or dry erase board; markers

🎯 GOALS FOR DAYS 2–5

Students practice using the bead-string model in the whole-group setting and count the number of beads to keep track of the number of days in school.

USING THE ROUTINE TO DEVELOP COUNTING STRATEGIES

1. Continue moving one bead for each day. Ask the following questions: "Why is it important that we keep track of the number of days?" and "What are we having on the one-hundredth day of school?" to help students remember why they are using the bead string and why it is important to keep an accurate record of the number of days ("We don't want to miss our one-hundredth day celebration!").

2. Offer opportunities for students to use and discuss a counting on strategy by asking, "We've been in school for four days. If I move over a bead for today, how many days will we be in school now?" Use this question *before* a student physically moves a bead. Expect students to be in different places with their counting strategies. They will have plenty of opportunities throughout the year to develop the counting-on strategy.

DAY 6 TO DAY 9: INTRODUCING THE NUMBER LINE MODEL TO ONE HUNDRED

FIGURE 2.4 On Day 6 the teacher introduces the growing number line and students begin to compare the bead string and the number line.

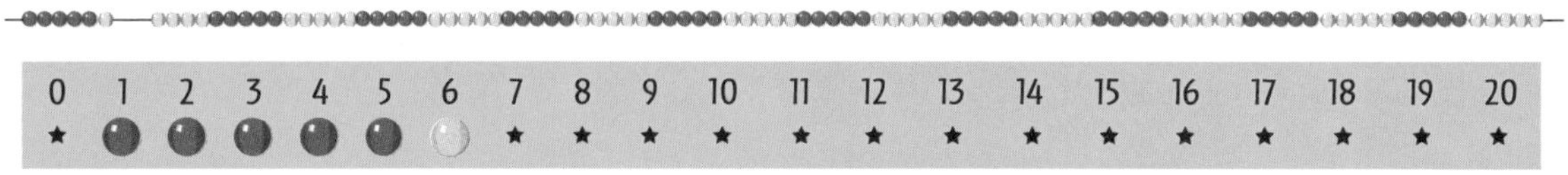

MATERIALS

- Bead string to one hundred
- Number line to one hundred (You can purchase one or make your own; see **Figure 2.4**.)
- Red and white paper beads (preferably laminated so that you don't have to remake these in the future) to keep track of the number of days in school on the number line
- Chart paper or dry erase board; markers

 ## GOALS FOR DAYS 6–9

Students continue to practice using the bead string model to figure out how many days they have been in school. They begin to make a connection between the bead string and the growing number of beads on the number line. For example, a child might say, "There are six beads pulled over on the bead string. Five are red and one is white. I noticed that there are five red beads on the number line and one white bead, too. It's the same."

Student Noticings About the Number Line

As students come to the meeting area, they will likely be excited to see this new model. Be sure to listen in to their comments, as these can give you insight into their mathematical understanding. Possible comments might be categorized in different ways:

- **Nonspecific:** "There are a lot of numbers."
- **Specific (naming numbers):** "I see the number 100! I see the number 87."
- **Counting:** "It goes 66, 67, 68, 69, 70, 71, 72 . . . "
- **Finding and naming patterns:** "All these numbers have 2's in them!"
- **Understanding the structure of the number system:** "There are tens—10, 20, 30, 40—and they go all the way to 100."

INTRODUCING AND USING THE NUMBER LINE MODEL

Have the number line up before starting the activity. If possible, place it so that it is eye level with the height of the children (Biddle et al. 2014).

1. Begin by naming children's noticings and excitement about the new model (e.g., "I see that many of you are excited by our new model. I heard Rashid say, 'There are so many numbers!'"). Say, "This model is called a number line, and we're also going to be using it to help us keep track of the number of days in school. Let's start by sharing what you noticed." (As students share out, record their noticings on a board or chart paper.) Say, "We're going to come back to this model in a minute after we answer the question 'How many days have we been in school?'"

2. After students answer the question "How many days have we been in school?" (they should have moved over the sixth bead on the bead string), pull out the paper beads that will be used with the number line. Show children the two colors and say, "We are going to be keeping track with these paper beads on our number line." Demonstrate how a bead will go below or above (or on) each number on the number line. Once you have put up the sixth bead, ask students, "What do you notice?" Students might compare the number line to the bead string.

POSSIBLE RESPONSES TO RECORD ON CHART PAPER

- "It looks like the bead string."
- "The number 6 has a white circle above it."
- "It goes red, white too, just like the bead string."
- "Five beads are red and five numbers have red circles [beads] above them."
- "On the bead string we have five red beads and one white bead; on the number line it's the same."

3. Another option for using the number line in this routine is to place the paper beads on top of the numbers as you count the days of school. This choice makes the number line more abstract because some numbers are no longer visible. It also nudges children to develop a way of thinking about the question "What number comes right before . . . ?" and helps them start to build a mental representation of the number system using existing patterns (e.g., before every multiple of ten there is a number with a nine in the ones place; after every multiple of ten is a number with a one in the ones place). Although most young students will not naturally use language like "ones place" (and we don't encourage teaching this language too early), they will be noticing and using patterns. Although the patterns students initially notice may seem simple in nature, celebrate their noticings! The first step in cracking the base ten number system is noticing that there are patterns (Valeras and Becker 1997).

Frankie's Journey: Day 10

It's the tenth day of school. Ten beads are pulled over on the bead string; the growing number line shows ten dots (alternating red and white groups of five) below the numbers. Children are seated in a horseshoe facing the board where the routines are displayed.

Classroom Dialogue

1. **Teacher:** We're here in school a new day. How many days have we been in school? Frankie?

2. **Frankie:** Seven.

3. **Teacher:** Hmm. Can you come up and show us the seven?

4. [Frankie stays seated; he is still reluctant to come up to the board in front of the room.]

5. **Teacher** [encouragingly]**:** Come on up.

6. [Frankie comes to the board where the bead string is, but does not know where to look.]

7. **Teacher:** Show us how you figured that out—how many days we've been in school.

8. [Frankie moves closer to the bead string and stares, but is reluctant to touch the beads. Other students call out at this point to help him; Frankie turns to look at them and smiles nervously.]

9. **Teacher:** You know, children, eyes are on Frankie because Frankie is helping us to figure this out now. Eyes on him, please.

10. [Frankie points to the bead string.]

11. **Teacher:** Can you show us how many days we've been in school?

12. **Frankie** [counts slowly, touching each bead]**:** One, 2, 3, 4, 5, 6, 7, 8, 9, 10.

13. **Teacher:** And today is a new day?

14. [Frankie nods yes.]

15. **Student** [calling out]**:** That makes it eleven.

16. **Student:** It's eleven.

(continues)

(continued)

Frankie's Journey: Day 10

17. **Teacher** [goes to bead string to assist him]**:** And so today is a new day? [Teacher moves the beads to demonstrate it's okay to touch the beads to help figure out this question.]

18. [Frankie nods yes again.]

19. **Teacher:** So, we were in school ten days and today is a new day. How many days is that?

20. **Frankie:** Eleven.

21. **Teacher:** Eleven. Are you sure?

22. [Frankie remains silent]

23. **Teacher:** Are you sure?

24. [Frankie nods yes.]

25. **Teacher:** Do you want to check? Go ahead.

26. **Frankie** [counting slowly and touching each bead on the bead string]**:** One, 2, 3, 4, 5, 6, 7, 8, 9, 10, 11.

27. **Teacher:** So, do we agree that we've been in school eleven days?

28. **Students** [in unison]**:** Yes.

Analysis of Teaching Moves

Several teacher moves in this vignette support and empower student thinking. First, the teacher encourages Frankie to share his thinking and does not indicate that his answer is incorrect [lines 1–3]. She invites him to the front of the room and notices that he is unsure of himself and may not feel comfortable speaking in front of his peers [lines 3–7]. Notice how the teacher slows down around his thinking and has the children pay attention to what Frankie says and does. All of these moves are essential in developing a community where each learner is respected and where the emphasis is not on the answer, but the communication of the thinking and the strategy for finding the answer [lines 9–18]. The teacher *does not* shy away from challenging Frankie with her question, "Are you sure?" This move helps children understand that math is not just about getting an answer, but about justifying *how they got their answer*.

Frankie's Journey: Day 10

Finally, this vignette shows that the routine can accommodate a wide range of learners. Although some children are already counting on (they know that if you add one more day to the ten days, it makes eleven days [line 15]), other children, like Frankie, are learning to count by ones and need to count each bead to know the total amount. The routines are also a way to establish classroom norms (what is valued and what is accepted).

Key Ideas of the Vignette

- Ownership of learning belongs to each learner.
- Each learner is respected no matter where they fall on the mathematical landscape.
- It is important to slow down around students' sharing of ideas.
- It is important to have students justify their thinking.
- Since the emphasis is on the process, not the product, it's okay for learners to change their answers.
- Models are helpful tools to find and justify thinking.
- The emphasis is not on speed, but on sense making. The community gives each other thinking space to understand ideas.

DAY 10: INTRODUCING THE TEN-FRAME

FIGURE 2.5 The *How Many Days Have We Been in School?*
routine on Day 10 of school

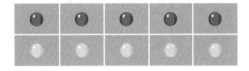

MATERIALS

- Bead string to one hundred
- Number line to one hundred
- Red and white paper beads to keep track of the number of days in school on the number line
- Blank ten-frames (the ten-frames should be copied on nonwhite paper so that the red and white dots that will be placed on it will be clear)
- Red and white dot stickers to match the colors used on the bead string and number line
- Chart paper or dry erase board; markers

🎯 GOALS FOR DAYS 10 AND BEYOND

Students continue to practice using the bead string and number line models to figure out how many days they have been in school. A new model, the ten-frame, which is structured in two rows of five, is introduced on Day 10 (see **Figure 2.5**). This model will give students more opportunities to reason with the five- and ten-structure. Although the organization of the ten-frame is similar to the bead string in the emphasis of ten as a unit that is comprised of two fives, it is different in that is a noncontinuous model. With the ten-frame, it may be easier for students to develop part-whole relations within ten, which is important for developing automaticity with the combinations of ten (Wheatley 1999).

INTRODUCING AND USING THE TEN-FRAME

1. Say, "Today I'm going to introduce a new tool [model] to help us keep track of how many days we've been in school." Show the children the ten-frame and ask, "What do you notice?" Chart children's responses.

    - "There are squares."
    - "It looks like a box."
    - "There are ten boxes."
    - "I see five and five."
    - "There are boxes on the top and boxes on the bottom."

2. Show the students the red and white dot stickers and say, "We are going to put these sticker dots on our ten-frame to keep track of how many days we've been in school. We'll put the red dots on the top row and the white dots on the bottom row. We've just figured out that we've been in school for ten days. I wonder how many red dots and how many white dots I'll need to show ten days."

    POSSIBLE RESPONSES
    - "Ten dots."
    - "Ten red dots and ten white dots."
    - "Five red dots for the top and five white dots for the bottom."

3. Have a student come up and put the red dots on the top row. Ask the students, "How many dots is that?" Even though you eventually want students to subitize (instantly recognize quantity) the five dots on the top row as a group, it may be helpful initially to have a student count to check the quantity of dots (Clements and Sarama 2014). It is still very early in the year; if a child says, "It's five dots—I just know," they may need to be reassured that it *is actually five*. After the students are sure there are five red dots, ask, "If there are five red dots on the top, how many white dots will I need to put on the bottom?" Accept all possible responses without commenting on whether answers are right or wrong (Boaler 2015). Have a child come up and put the white dots on the bottom row of the ten-frame and check to answer the "how many" question (e.g., "How many white dots is that?"). Say, "We have five red dots on the top and five white dots on the bottom. That's ten dots in all."

Teacher Note

Developing Big Mathematical Ideas over Time

It is often challenging for students who have not yet developed cardinality (the idea that the last dot counted represents all of the dots, the set of ten, in this case) to answer a question that requires them to use part-whole relations (Kamii 1985). It seems obvious to us as adults that if there are ten total squares on the ten-frame, you cannot have ten red dots and ten white dots because that would give you more dots than squares. However, for students who are still developing important ideas about quantity and how the number system works, this is a big idea, one that will take time to develop. It is important that when students give incorrect answers, we do not try to "fix" their thinking in the moment (Steuer, Rosentritt-Brunn, and Dresel 2013). Students need time and opportunities to grapple with these ideas, which is a primary purpose of this routine. This routine helps students develop their mathematical reasoning because it gives them the opportunity to try out their ideas and, if the ideas are incorrect, adjust how they think. Some students will need more time than others and that's okay. Remember, learning is a journey and not everyone is on the same path at the same time!

Frankie's Journey: Day 24

Classroom Dialogue

1. **Teacher:** How many days have we been in school? How would you like to figure that out?

2. **Frankie** [goes to the ten-frames and starts counting from the bottom dot on the first ten-frame]**:** One, 2, 3, 4 . . . 24.

3. **Teacher:** How many days have we been in school?

4. **Frankie:** Twenty-four days.

5. **Teacher:** I noticed Frankie did something different today. Did anyone else notice what Frankie did?

6. **Mia:** He started in the wrong place. He didn't start with one.

7. **Frankie:** I said, "One."

8. **Mia:** But you pointed down there [points to the last dot on the bottom row in the ten-frame], and you need to point here [indicates the first dot in the first row of the ten-frame]. This is one.

9. **Teacher:** Who understands what Mia just said? Ahmed?

10. **Ahmed:** Frankie made a mistake where he started. Can I come up? [He comes up to the ten-frame chart.] This is one [points to the first dot on the top row of the ten-frame]. Frankie did this [points to the last dot in the second row on the ten-frame]. He said, "One," and that's not one, it's ten.

11. **Teacher:** Hmm. It sounds like we have a disagreement here. [The teacher addresses the class.] Does it matter where we start counting on the ten-frame? Can I start counting from here [shows where Frankie started] or do I have to start here? Mia and Ahmed think we have to start here [points to the first dot in the first row of the ten-frame]. Do you agree with Frankie or with Mia and Ahmed? Turn and talk to your partner about this question. When we're trying to figure out how many days we've been in school, can we start counting anywhere on the ten-frame or do we have to start here [points to the first dot again]?

(continues)

(continued)

Frankie's Journey: Day 24

Analysis of Teaching Moves

This dialogue represents a big moment for children's mathematical development, one that the teacher capitalizes on by slowing down the conversation and bringing in different voices ("Do you agree with Frankie or with Mia and Ahmed?") and having children talk to a partner about what happened. Does it matter where we start counting on a ten-frame? Although this may seem like an obvious question to adults, it is often puzzling to students who think that you have to start with the first dot (the child says, "This is one"). These students conflate position (the first dot in the first row) with quantity (one dot is one dot no matter where you start counting).

Students are grappling with these big math ideas:

- Order doesn't matter when you are counting. What matters is that you organize what is counted in a way that helps you remember the starting and ending place, ensuring that everything is counted (in this instance, the dots on the ten-frame).
- When counting, the number said is not a label, but a quantity. This idea is a big developmental leap for children as they develop counting strategies. For example, when Frankie says "Two," he is referring to the two dots counted, not to the second dot. The development of the idea of set is critical to the development of part-whole relations (that any quantity can be decomposed and recomposed) and that the sum of the parts is equivalent to the whole.

DAY 30: INTRODUCE THE HUNDREDS CHART

FIGURE 2.6 Children explore the hundreds chart as part of the *How Many Days Have We Been in School?* routine on Day 30.

1	2	3	4	5				9	10

The hundreds chart can be introduced around the thirtieth day of school, once students have had the opportunity to develop familiarity with the bead string, number line, and hundreds chart as models. We recommend using a pocket hundreds chart, which can be ordered from most math education supply companies.

MATERIALS

- All previously used tools/models
- Pocket hundreds chart (cards that can be moved, turned around, etc.; see **Figure 2.6**.)
- Chart paper or dry erase board; markers

🎯 GOALS FOR ROUTINE

By this point, most students will be very familiar with the *How Many Days Have We Been in School?* routine. This new model is introduced to further support the sequencing of numbers in base ten and to help students think about the patterns that occur in our number system (e.g., in a column on the hundreds chart, *all* the numbers have a three in the ones place). Note that the structure of the hundreds chart is not always immediately clear to students (e.g., why is ten at the end of one row and eleven is not beneath it, but all the way to the left?). As with the other models, students need time to explore the hundreds chart and develop an understanding of how it works (Conklin and Sheffield 2012).

INTRODUCING AND USING THE HUNDREDS CHART

1. Begin with no number cards in the hundreds pocket chart. Say, "Here's another new tool we're going to be using to keep track of how many days we've been in school. What do you notice?" Chart children's responses.

POSSIBLE RESPONSES

- "It's a big box."
- "It's a big box with little boxes."
- "It's got plastic on it."
- "It looks like something goes in there [the pockets]."

2. Say, "I'm going to do something now and I want you to watch and notice what I'm doing. Think in your head and then we'll talk when I'm done." Begin to put the number cards in the first row (1, 2, 3, 4, 5, 9, and 10), but *do not* fill in all the numbers on the first row.

3. Ask students, "What do you notice now?"

POSSIBLE RESPONSES

- "There are numbers."
- "It's counting, 1, 2, 3, 4, 5."
- "Numbers are missing. Six is missing."
- "You don't count like that, 1, 2, 3, 4, 5 . . . Where's 6?"
- "I think the ten [tens] will go here [in the right column]."
- "Eight will go before 9; 6 will come after 5."

4. Use students' comments to help you place the cards in the first row of the hundreds chart. For example, if a child says, "Six comes after five," highlight that idea and then add *6* to the chart. Using student predictions helps them think about patterns in the number system and how they can use these patterns to learn what number comes before or after another number. The hundreds chart is the perfect model for students to start to notice and reason with these patterns.

5. Once you have explored children's conjectures and put some of the cards on the hundreds chart, say, "We used our other models this morning to figure out that we've been in school for thirty days. We are going to use the hundreds chart as a tool to help us as well." Choose a number that isn't yet on the hundreds chart (e.g., fourteen). Ask students, "Where would we place 14?" After students come to an agreement, place *14* on the chart.

6. Give pairs of students the remaining cards up to 30 and have them place the numbers on the chart.

Frankie's Journey: Day 37

Classroom Dialogue

1. **Teacher:** How many days have we been in school? Frankie?

2. **Frankie:** [goes to the ten-frames and counts] Ten, eleven, twelve . . . thirty-seven.

3. **Teacher** [in enthusiastic voice]**:** So, you know what I noticed? [The teacher looks at the class with surprise on her face and whispers.] Frankie didn't have to count all of these [indicates the dots on the first ten-frame].

4. [Frankie nods in agreement.]

5. **Teacher:** Why not?

6. **Frankie:** 'Cause it's ten. All of these is ten [indicates all the dots on the ten-frame].

7. **Teacher** [to the class]**:** Did you hear that?!

8. **Student:** He didn't count 1, 2, 3, 4, 5, 6, 7, 8, 9, 10. He knew it was ten!

9. **Teacher:** You know what, Frankie? I want to know if anybody else saw what you did and can come up and show us.

10. **Student** [comes up and points to first ten-frame]**:** He touched this and said, 'Ten.' Then he said, 'Eleven, twelve, thirteen . . .'

Analysis of Teaching Moves

Effective teaching is directly connected to a teacher's ability to notice (and listen to!) big ideas in children's strategies and to capitalize on those moments to deepen the thinking of the community. What does this vignette reveal about what the teacher notices and values? She

- recognizes that Frankie has made a major shift in his counting strategy (he counts on in line 2);
- knows that because many of her children are still counting by ones, this is a moment worth highlighting and celebrating;

(continues)

(continued)

Frankie's Journey: Day 37

- is willing to slow down the learning around big ideas (a child cannot use a strategy like counting on without having developed the idea of set—ten is a set and "I don't have to count it to know it's ten");
- has Frankie explain *why* he doesn't have to count the dots on the ten-frame;
- involves other children in the strategy by asking them to repeat Frankie's strategy;
- finally, slows down to explore Frankie's strategy, and uses her voice—whispering, "Did you see what he did?!" to share her wonderment.

What better way to empower Frankie (and all the students in her classroom) than to celebrate the power of ideas!

BEYOND DAY 30

FIGURE 2.7 The *How Many Days Have We Been in School?* routine on Day 31 of school

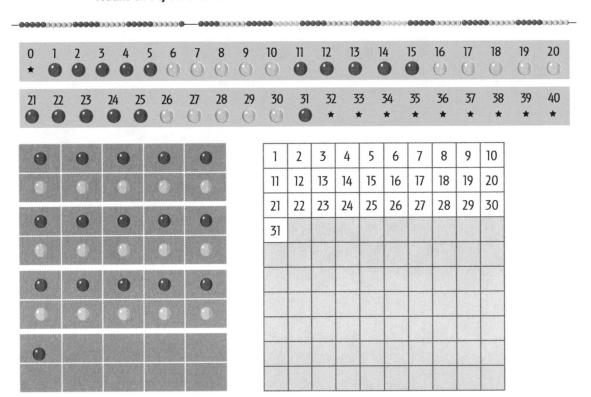

When children are familiar with all the models in the *How Many Days Have We Been in School?* routine (see **Figure 2.7**), the routine no longer needs to be done during the whole-class morning meeting. In fact, it is important to shift the ownership of the routine to the children so that they have more direct experiences with the models and opportunities to justifying their thinking to classmates (Crowe and Kennedy 2018). Each week a different pair of children can be responsible for updating the *How Many Days Have We Been in School?* models daily. On some days the teacher may choose to interact with the pair of children as they are working. The teacher focuses on asking questions that develop students' thinking rather than on fixing inconsistencies. For example, if the teacher notices that the bead string shows fifty-six days while the hundreds chart shows fifty-eight, she might simply say, "So, how many days have we been in school?" rather than "The bead string and the hundreds chart don't match." Remember, the goal is to develop students' ownership and reasoning with the routine, and mistakes are prime opportunities for this learning.

Over time, the teacher's role shifts. Children begin to notice discrepancies between the models without prompting. The teacher's role then becomes to provide space for students to debate their ideas so that children's questions and comments stay front and center. On some days, the teacher may choose to devote some time during morning meeting to discuss children's ideas, questions, and mathematical disagreements.

Frankie's Journey: Day 56

Classroom Dialogue

1. **Teacher:** Do you want to come up and use the bead string to show us how many days we've been in school?

2. **Frankie** [counts by tens, pulling over two groups of five as he does so]: Ten, 20, 30, 40, 50 . . . [pauses] 50 . . . [is unsure how to count on from 50; stares at the remaining beads].

3. **Teacher:** Who can say what they saw Frankie do?

4. **Student:** He counted by tens. He touched the ten each time. He said, "Ten, 20, 30, 40, 50."

5. **Student:** We've been in school fifty-seven days.

(continues)

(continued)

Frankie's Journey: Day 56

6. **Teacher:** How can you help us figure this out? Is it fifty days or fifty-seven days?

7. **Student** [comes up and touches the beads as he counts]**:** This is 50—Frankie counted these. Then you say, 51, 52, 53, 54, 55, 56, 57.

8. **Teacher:** What do you think, Frankie? Do you agree or disagree?

9. **Frankie:** Agree.

10. **Teacher:** Why do you agree?

11. **Frankie:** 'Cause there's more beads here [indicates the seven beads he didn't count]. I didn't count these.

12. **Teacher:** Would you like to try?

13. **Frankie** [touching groups of ten as he counts]**:** Ten, 20, 30, 40, 50 [long pause] 50 . . . 51, 52, 53, 54, 55, 56, 57.

14. **Teacher:** How many days have we been in school?

15. **Frankie and other students:** Fifty-seven.

Analysis of Teaching Moves

This routine allows every student to develop at their own pace. Frankie has moved beyond counting by ones and is now skip-counting by tens. However, even though he is using the structure to count by bigger units (e.g., ten), he is still learning how to shift from one unit to another (e.g., he's counted to fifty by tens, but has to shift back to counting by ones). This shifting back and forth between units is often difficult for young children, which is why we often see them counting the fifty-first bead as "sixty." Being able to reason with different units (unitizing) is the heart of place-value understanding, and Frankie is starting his journey to make sense of how our base ten number system works.

The teacher use several moves to capitalize on the big ideas and create disequilibrium. First, she recognizes the shift in Frankie's counting strategy and slows down to highlight this change with the class [line 3]. Second, she signals that the other students need to be making sense of what is happening [line 3] and invites them to share their noticings. Third, she knows Frankie is still learning to count on from fifty and allows disagreement to arise when another child says, "We've been in school

Frankie's Journey: Day 56

fifty-seven days." The teacher asks (and remains neutral in the asking), "Is it fifty days or fifty-seven days?." Finally, when Frankie changes his mind, she asks him why and invites him to try a strategy that has just been shared. She also uses wait time to give him thinking space.

In this classroom, the emphasis on thinking and communication sets high standards for all children's mathematical learning. The teacher's expectation that Frankie can justify his thinking opens the door for him to think about his strategy and realize that he "didn't count these." Her emphasis on thinking also removes any onus about making mistakes—which are a natural and important part of learning.

The Power of One Routine in Supporting Student Reasoning

Throughout this chapter, you have read about the *How Many Days Have We Been in School?* routine and followed Frankie's journey as a mathematical thinker. Although the effect this routine (and the teaching that supported him) had on Frankie's thinking can be traced in the transcripts, an even more powerful testimonial came at the end of the year, when Frankie's mother wrote a letter to his teacher. Here's the teacher's summary of what Frankie's mother said:

> " At the end of this year I shared the video clips and write-up of Frankie's journey with his mother, Maria. She began by telling me that his journey was about more than just an understanding of math and numbers. It was through the development of these routines that Frankie grew emotionally. She had great difficulty with him at the beginning of the year. She struggled getting him up in the morning and to school on time without encountering a daily battle. Maria realized that as Frankie was given more responsibility for the math routines in class, the battles became less frequent and his excitement for school grew. Frankie started to wake up with a smile on his face and was ready to come to school to learn. His response to the question "How was your day?" went from him saying nothing to him initiating the conversation about all the things he learned in school and specifically about math. She was overjoyed watching his journey and grateful that he felt empowered in his own learning. "

Playing with this routine was powerful for us as teachers. It gave us new insights into our teaching practices and the children we teach. We found that as we stepped back, children stepped forward. As we grew, they grew. Frankie, and other students in our classrooms, taught us that when students are challenged to engage with mathematical big ideas through routines, they rise to the challenge, leading the way with their noticings, questions, and thinking.

Making the *Attendance* Routine Matter

> " Children must be taught how to think, not what to think. "

Margaret Mead, anthropologist, *Coming of Age in Samoa*, 1928

Attendance as a Rich Context for Developing Mathematical Thinking

"How many children are here today?" is a commonly asked question in early childhood classrooms, often used to launch a daily *Attendance* routine. And although, to many adults, this question may seem easy enough to answer (just count the people!), early childhood teachers know that figuring out the answer to this question can be quite challenging for young children. These productive challenges, however, may arise only if we, as teachers, resist the temptation to control the situation—show kids how to count, how to organize themselves to be counted, or correct their sequencing errors ("What comes after fourteen?"). Young children may work on the "How many children are here today?" question for days or even weeks before finding a system to answer this question accurately.

And yet, too often we do not give students the time or space to productively struggle with this question. Perhaps it is because we feel constrained by time ("Kids have too much to learn to spend all this time counting—especially if I can help them!") or because we're uncomfortable watching students struggle or make mistakes publicly. The good news is that we can reimagine the traditional *Attendance* routine in a way that allows children to develop critical counting and computation strategies, key communication skills, and the self-confidence that comes with true problem solving.

Who Owns the Thinking? Who Owns the Math?

The most critical question to consider as you read the kindergarten vignettes in the following section is "Who owns the math?" (This is a question one of Toni's teachers, Maarten Dolk, would ask at the Math in the City project when she was a new teacher first coming into contact with novel ways to teach mathematics.) One way to think about this question is to picture yourself as a learner in each classroom and think about these two questions: (1) What are you learning mathematically? And (2) What are you learning about your role as a learner? It is important to remember that as we do routines with children, we are teaching them more than math content. We are teaching students what it means to learn and what it means to be a member of a community of thinkers (Ergas 2017). How we use our routines is an enculturation process. The learning community we create as teachers is directly connected to our beliefs about the role of the student and our role as the teacher. The kinds of questions we ask, when we ask them, and why we're asking them send powerful messages to our learners (Tishman, Perkins, and Jay 1993).

The *Attendance* Routine in Kindergarten Classrooms

Let's take a peek into two different kindergarten classrooms to think about why the question "How many children are here today?" is challenging for many young children. As part of our analysis, let's also think about how the role of the teacher either opens the door for thinking and productive struggle or closes it (Dweck 2006).

Here are two questions to think about as you compare these classrooms: (1) What are the goals of the *Attendance* routine in each classroom? (2) What are the teacher beliefs about (a) the *role of the routine* in developing reasoning; (b) the *role of the students* in learning; and (c) the *role of the teacher* in supporting that learning?

Kindergarten Classrooms 1 & 2

Kindergarten Classroom 1

1. **Teacher:** How many children are here today? Who remembers what we did yesterday?

2. **Child:** We count.

3. **Teacher:** We counted. What did we count?

4. **Child:** Kids.

5. **Teacher:** We counted the kids. Great. And what did we do to help us count the kids? José?

6. **José:** Kids stand.

7. **Teacher:** Kids stood up and then what happened?

8. **José:** We touch them.

9. **Teacher:** We touched each child and then what happened?

10. **José:** They sit down.

11. **Teacher:** Okay, let's try that. José, why don't you count?

12. **José** [touches each child as he counts]**:** One, 2, 3, 4, 5, 6, 7, 8, 9, 10, 15 . . .

13. **Teacher:** Uh-oh! Wait. This was ten [touches the tenth child]. This next one would be [touches the eleventh child] . . .

14. **José:** Fifteen.

15. **Teacher:** Eleven. This would be [touches the twelfth child] . . .

16. **José:** Fifteen!

17. **Teacher** [corrects him quickly]**:** Twelve.

18. **José** [looks at teacher and repeats]**:** Twelve. Fifteen [in uncertain voice as he touches the next child]?

19. **Teacher:** Thirteen.

20. **José** [repeats]**:** Thirteen. [He moves to the next child.] Fifteen?

21. **Teacher:** Fourteen.

(continues)

(continued)

Kindergarten Classroom 1

22. **José** [repeats]: Fourteen [looks to the teacher as he touches the next child's head, but does not say anything].

23. **Teacher:** Fifteen.

24. **José:** Fifteen. [The count continues in this fashion with the José touching each child and the teacher providing the number name until he counts all twenty-four children.]

Analysis of Teaching Moves

In line 1, the teacher asks a genuine question, which is challenging for many of her students to answer (it presents a real problem). However, her next question, "Who remembers what we did yesterday?," robs the first question of its power. Although the intention behind asking the second question has the potential to help children count more accurately, it also takes away their ownership. What if some children have their own system for figuring out how many children are in the classroom or haven't yet developed the need for organizing? What if the children don't remember what happened yesterday? Does that mean they wouldn't be able to answer the first question?

Changing the question in line 3 to "How did we count yesterday? Can you show us?" and allowing the child to try it out would quickly have let the teacher know whether this child understood the standing-up system previously used by another child.

Notice that the teacher questions in lines 1–11 are about helping the student remember all the things that happened in the order in which they occurred.

In line 12, José does not remember what comes after ten and says fifteen. Because the teacher is focused on the act of counting (getting the answer to the question "How many children are here today?"), she corrects him. After this correction, José looks to his teacher for affirmation of his answer and eventually becomes hesitant to give an answer. A different approach would be to let him finish his count without correcting him and have the wrong answer. What might be the benefits of this approach?

Kindergarten Classroom 2

Kindergarten Classroom 2

1. **Teacher:** I bought these new folders for us to use today [shows the folders]. And I want to give one folder to each kid who is here. I'm wondering how many children are here today. Who thinks they can help us answer this question? Alex?

2. **Alex** [gets up, walks around tagging and accurately counting some kids, but does not tag and count all of the children]: One, 2, 3, 5 . . . 15. Fifteen kids.

3. **Clyde:** You forgot to count yourself. So, it's sixteen.

4. **Josh:** It's more than sixteen. Alex missed kids [points to a group of kids]. He didn't count them!

5. **Ami** [to Alex, helpfully]: Look, you didn't count right. You have to count all the kids; like this [touches kids as she counts]: 1, 2, 3, 4 . . . 22, 23. See? Twenty-three kids. I counted everyone. You counted like this [indicates he was just counting in the air and not touching the kids].

6. **Teacher:** We seem to be having some disagreements. Alex counted fifteen kids and Clyde said he forgot to count himself, so it's sixteen kids [writes *15* and *16* on the board]. Josh said it's got to be more than sixteen kids. Then Ami counted twenty-three kids. I'm wondering, does it matter if we get different answers when we count the kids?

7. **Students** [answer at the same time]: Yes. No.

8. **Teacher:** I hear some yeses and some noes. Some of us think it matters that we get different numbers of children when we count and some of us don't think it matters. What should we do? Sue?

9. **Sue:** Count again [gets up and counts, missing some students and some numbers in the counting sequence]: One, 2, 3 . . . 11, 12, 13, 14, 16, 17, 12, 14, 18. Eighteen kids.

10. **Jiro:** You missed some numbers and some kids. It goes fourteen, fifteen—not fourteen, sixteen!

11. **Teacher** [writes *18* on the board]: Oh, my! This is so challenging. Are there fifteen children here today? Sixteen children? Eighteen? Twenty-three? Does it matter if we get different counts? Jody?

12. **Jody:** Sometimes when you count you get different answers.

(continues)

(continued)

Kindergarten Classroom 2

13. **Teacher:** So, you're expecting to get different answers when you count?

14. [Jody nods her head.]

15. **Manny:** How can it be? The kids are the kids!

16. **Teacher:** What do you mean, "The kids are the kids"?

17. **Manny:** We're all here. It's not more kids, not less kids. Just kids. It should be the same.

18. **Teacher:** Does someone understand what Manny is saying? Can you put it in your own words?

19. **Students:** No.

20. **Jiro:** He says no one left the room so how can we have less kids? No one came in, so how can we have more? It's gotta be the same when we count.

21. **Teacher:** This is so challenging! We'll keep working on it. I do want to give out our folders though. Sue, you said there were eighteen kids here today. How many folders will we need for eighteen kids?

22. **Sue:** Eighteen.

23. **Jiro:** That's not enough.

24. **Teacher:** Let's count them out together. [The class counts the folders to eighteen.] Could you give them out to the students please?

25. [Sue gives out the folders to the students, and comes back to the teacher, perplexed.]

26. **Manny:** Hey! I didn't get a folder!

27. **Sue:** We need more.

28. **Teacher:** What do you mean, "We need more"?

29. **Sue:** It wasn't enough.

30. **Teacher:** I'm confused—I thought there were eighteen children here today.

31. **Sue:** There's more.

32. **Teacher:** More than eighteen kids?

33. **Sue:** Yes.

34. **Teacher:** I wonder how many kids are here today. I'm wondering if there's a way for us to figure this out. We'll think about this question some more tomorrow. For now, anyone without a folder please come up and get one.

Kindergarten Classroom 2

Analysis of Teaching Moves

- The teacher poses a question and lets kids grapple with it.
- She is okay with wrong answers.
- She writes all kids' counts on the board and doesn't say which one is correct. The ability to remain neutral and not signal the answer puts the ball in the kids' court—they have to figure it out!
- She enters into the conversation to pose big idea questions: "Does it matter if we get different answers when we count the kids?" [line 6].
- She's okay with ambiguity: Is it fifteen? Twenty-three? Eighteen? This ambiguity creates space for students to have productive struggle. As part of this struggle, students offer their own solutions to what's problematic.
- Enters in the conversation once again to pose big idea questions, such as "What do you mean, 'We need more'?"
- Doesn't wrap the lesson up with a bow: "I'm wondering if there's a way for us to figure this out. We'll think some more about this question tomorrow." [line 34].

Comparing the Teaching in Two Classrooms

The question "How many children are here today?" becomes an authentic problem in Classroom 2 because it is directly linked to the need for an accurate count (we want everyone to get a folder, and when we don't have an accurate count, someone may not get one). The fact that there is a problem with the count isn't immediately apparent to the children—it's not until all the folders are handed out that the purpose of counting becomes apparent. As Manny says, "Hey, I didn't get a folder!"

Although the teachers in both classrooms are guiding the learning around students' mistakes, their intentions appear to be different. In Classroom 1, the teacher is trying to fix the counting mistakes; in Classroom 2, the mistakes are used as an opportunity for kids to grapple with whether it matters if they get different counts. In this classroom, notice that it is the children who correct the mistakes. As students watch each other count, they begin to offer their own ideas about how to get an accurate count of the students. They are correcting each other and building on each other's thinking.

Notice that the teacher in Classroom 2 sits back a bit and in doing so empowers her students to do the work. The students speak to each other without going through the teacher. The teacher seems to know when to enter into the conversation (around big ideas) and when to let children play with ideas. Notice that when the teacher steps back, the students step up.

Figuring out how many kids are in class is a challenge for many young students. In both classrooms we see students struggling in different ways with this question. How the teachers respond to children's struggles, however, is different. One major difference is reflected in the pedagogy (Schon 1983). In Classroom 1 the teacher overscaffolds (she supplies the numbers that are not known in the counting sequence) and in doing so undermines the child's confidence. In Classroom 2 the teacher lets the children fail but that failure leads to high engagement and ownership of the problem. Although students may not have gotten the correct count in Classroom 2, they are acutely aware of the fact that there is something awry. Because their struggles are not immediately fixed by the teacher, they will have to continue to grapple with their counting until they come up with a system for keeping track.

Furthermore, in Classroom 2, the teacher makes the *Attendance* routines purposeful. It's not just *school math*—we don't take attendance just to take attendance—we really need to know how many kids are here so that we have enough pencils, folders, or milk for snack. The beauty of a real context, in which the teacher gives the students an opportunity to own the problem, is that the solution often isn't immediately apparent, and students must struggle to answer the question (Van den Heuvel-Panhuizen 2000). The productive struggle in Classroom 2 doesn't just develop students' mathematical thinking; it shapes their view of learning and sets expectations for the community as a whole. The struggle isn't that of one individual—it is a community struggle. The big difference between Classroom 2 and many other classrooms is that when the solution comes, the students *own* it.

Getting Started with the *Attendance* Routine

Here are some questions to consider with any routine you use in your classroom:

1. Why are you doing this routine? What are the learning goals for your students?

2. Why are you doing this routine in the way you're doing it? (For example, if you are passing out a cube to each child, then collecting the cubes and putting them together to represent the total number of children present, why are you structuring the routine in this particular way?)

3. What do your students know mathematically and how will this routine support their growth and development?

4. How will this routine, and the way you structure it, help your students develop their ability to communicate and become socially and emotionally aware members of our classroom community?

As you get started with the *Attendance* routine, it is important to note that this routine is designed to be introduced without a formal structure. There is great value in providing the time and space for children to figure out a way to answer the "How many?" question without using a specific tool or model chosen by the teacher. This kind of "figuring out" in the *Attendance* routine mirrors what happens in the real world when we are asked to answer the question "How many?"

Two teacher moves are important to develop as you get started with the *Attendance* routine. The first is to anticipate how children will grapple with the attendance problem; the second is to challenge each child to think, no matter where they are currently in their understanding of the problem (Smith and Stein 2011). Here are three ways children often think about the question "How many children are here today?" as they get started with the *Attendance* routine:

1. Some children have already developed counting strategies and can easily figure out how many children are present. Although counting is not problematic for them, these students can be challenged in other ways (e.g., how do they convince someone that their count is accurate? If they have a system for counting accurately, how do they communicate that to peers? How do they use what they know about counting to help someone else who is struggling? What do they say? How do they say it? What do they do if others don't listen to them or agree with them?).

2. Some children find the question "How many?" challenging. Why? Counting is rooted in children's development of some big ideas in early number (see Appendix B for a list of some of these ideas). Counting requires the coordination of four key things: (1) knowing the sequence of number names (the rote counting sequence); (2) tagging each object to be counted; (3) giving each object counted one name so that your voice is in sync with what you count; and (4) organizing your count so that you can keep track of what has and hasn't been counted.

3. Finally, some children do not yet realize that the question "How many?" is about quantifying (finding a number that names the quantity of the set counted). For them, it is a cue to sing the counting words. In fact, you may see these children counting the same set over and over (if you ask, "How many?" four times, they will recount the same set four times). They are not bothered if they get different counts each time because they do not yet understand that the number word signifies a quantity (or a set).

Children who are not yet aware that counting is about *quantifying* may not understand other children's attempts to solve the problem at first. One way to support these children and all children is to use talk moves (e.g., pair talk, paraphrasing) to slow down around specific things that are happening: "Who noticed what Ami did when she counted the children? Turn and talk to your partner about that" (Chapin, O'Connor, and Anderson 2013). Another support is to increase the number of counting opportunities children have throughout the day.

We should also point out that the rote counting sequence itself is not easy to master. The logic of the number names in English breaks down after ten (Ng and Rao 2010). There are many counting words in English for young children to remember, including all the teens words and decade shifts (e.g., thirty, forty). Because the children's understanding of the counting system and number names may break down at certain points, you sometimes hear children struggle with certain numbers (e.g., children might say, "ten-one" for eleven or "ten-two" for twelve; they might call thirty "twenty-ten"). They may also have difficulty remembering the numbers fifteen and sixteen, which are sometimes skipped by young children when counting.

Teacher Note

Children's Organizational Strategies

Keeping track of people to be counted presents a real challenge to students. Be prepared for children to create different systems for organizing; they may have students stand before they're counted and sit when they've been counted; they may put each other in rows to be counted; they may count the girls first and then the boys. We once saw a child use the number line that was up in the classroom as part of the *Attendance* routine. He used one hand to point to the number on the number line and the other hand to point to a child being counted. Initially, children thought this was an amazing strategy and tried it out. Eventually, students realized that they were making mistakes (sometimes one hand moved slightly when it pointed to the number line, and the count was lost or inaccurate). The key point here is that the number line strategy was an invented strategy that was accepted and then abandoned by children when they realized it was problematic. To develop children's reasoning and ownership of the math, it's important to let them create and abandon their counting strategies without adult interference. Although this may initially feel difficult or unnatural to you, the payoff will come when you see how empowered students become by their own ideas. Children know when something is not working, and through debate, often abandon unproductive strategies. We have found that creating a real thinking space develops their ingenuity, flexibility, and persistence—all important problem-solving habits.

Introducing Structured Materials and Mathematical Models: Three Variations of the *Attendance* Routine

Because we value children's invented strategies (Freudenthal 1991) and know how powerful these can be for developing their ability to reason and communicate, we hold off on using structured materials and mathematical models. Structured materials (cubes, counters) and mathematical models (Rekenrek, bead string, ten-frame) can remove cognitive obstacles (such as the need for organization and a system for keeping track of the count) that are necessary for learning how to count. Once children have created systems for organizing their counting, then structured materials and mathematical models can be introduced as part of the *Attendance* routine. These tools will help students move beyond counting by ones and develop big mathematical ideas and place-value understanding. There is no exact schedule for when to introduce structured models; it all hinges on children's development (with some classes the structured models are introduced in the first or second week of school, in others it happens at a later date). Additionally, there is no one grade level that corresponds to each routine variation. A teacher should introduce the variation when students are ready to grapple with the specific math ideas that variation focuses on. (See **Figures 3.1**, **3.2**, and **3.3**.)

Attendance Routine Variation 1: The Arithmetic Rack Model

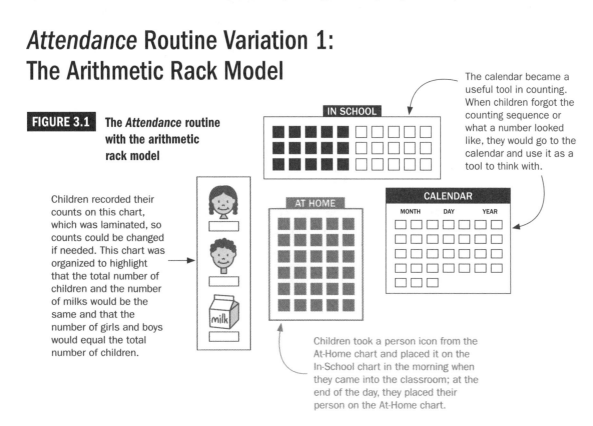

FIGURE 3.1 **The *Attendance* routine with the arithmetic rack model**

The calendar became a useful tool in counting. When children forgot the counting sequence or what a number looked like, they would go to the calendar and use it as a tool to think with.

Children recorded their counts on this chart, which was laminated, so counts could be changed if needed. This chart was organized to highlight that the total number of children and the number of milks would be the same and that the number of girls and boys would equal the total number of children.

Children took a person icon from the At-Home chart and placed it on the In-School chart in the morning when they came into the classroom; at the end of the day, they placed their person on the At-Home chart.

FIGURE 3.2 | An attendance chart structured like the arithmetic rack with rows of ten, five red and five white (Fosnot and Dolk 2001)

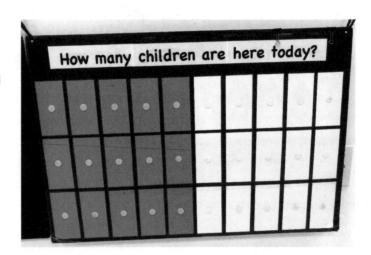

FIGURE 3.3 | The arithmetic rack or, as the Dutch call it, the Rekenrek (Treffers 1991)

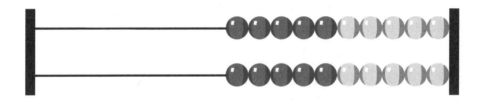

There are several reasons we use the arithmetic rack model in the *Attendance* routine. First, its organizational structure (alternating groups of five, highlighted by two different colors) emphasizing the importance of ten as a unit is a structure we use in other routines (e.g., *How Many Days Have We Been in School?*). Second, the arithmetic rack is a powerful model for developing part-whole relations and big ideas in base ten (e.g., 20 is two tens; 9 is 1 less than 10; 6 is 5 and 1). Third, the arithmetic rack becomes an important tool when students start developing fluency with the basic facts to twenty (see **Figure 3.3**). An example of how the structure of the arithmetic rack model can be used with the *Attendance* routine is shown in **Figure 3.2**.

MATERIALS

- Foam board
- People or face icons printed on cardstock in different colors and laminated
- Velcro to affix to the back of the people icons
- Basket for icons

ROUTINE STRUCTURE

As children come into the classroom in the morning, they take an icon from the At-Home basket or chart (**Figure 3.1**) and put it on the attendance board. The question "How many children are here today?" is used in conjunction with the board. Initially, this activity might be done with the whole class, and eventually it can be turned over to two students as one of the morning routine jobs. When this is done, the paired students would figure out how many children are present and report this information to the other children during morning meeting. As part of their presentation the pair of students might *prove* how they figured out the number of children present. Discrepancies are explored with the total during other parts of the day (e.g., during snack), when the final count is used to distribute food or napkins, and so on. The lack of accuracy (e.g., when a child says during snack, "Hey! I didn't get a milk!") makes children much more attentive to their counts—there is a real reason to be accurate!

To answer the question "How many children are here today?" in the *Attendance* routine, children may

- count icons by ones using the structure of the board (going across the rows) or count randomly (start anywhere on the board—count across, up, down, etc.).
- realize the first row will always have ten icons when it's filled and count on (e.g., point to the first row and say, "Ten," and then count the next row, "eleven, twelve, thirteen").
- skip-count by tens and count on (10, 20, 21, 22, 23, 24).
- skip-count by fives and count on (5, 10, 15, 20, 21, 22, 23, 24) or use the structure (5, 10, 15, 20, 25, count back one, 24).
- use the five structure to figure out sets less than ten. ("I know this is six children because this is five, and this is one more. Five and one is six.")
- start to work with part-whole relations and notice how many icons are left in the basket. If there are no icons left in the basket, they might say, "Everyone is here today. I know we have twenty-four kids in the class when everyone is here, so I don't have to count the kids—I know it's twenty-four!" Some children may extend this understanding to think about different amounts of cards left in the basket (e.g., "There are three cards in the basket. That means there are three people absent. If there are twenty-four kids in our class, you can count back three to find out how many kids are here today."). Although the last example is not a common strategy *initially* in preschool and kindergarten, kindergartners and first graders often develop this strategy as they gain experience with the routine.

Icons on the Attendance Board

Although some teachers use photos of their students' faces on the attendance board, we have found that photos can become problematic when some children want to rearrange the children's icons to show different ways of counting. Some children don't "want to be touched" by other children, and get upset. Sometimes children want to put their icons near their friends' and become territorial about this as well. We have found that using generic face icons avoids these issues.

USING THE *ATTENDANCE* ROUTINE TO MAXIMIZE STUDENT LEARNING

As you use the arithmetic rack model for the *Attendance* routine, two important things to remember are (1) do not tell children where to place their icons and (2) do not ask them to fill in one row before they start another. When children initially work with this model, you may see them randomly place their people icons on the board. Young children are often not yet aware of the attendance board's structure and how it can help them count in more efficient ways. Showing children what to do and how to do it robs them of this powerful aha moment ("If this row is ten, when it's filled, I don't have to count it because I just know it's ten!"). Over time, as children begin to recognize the structure of ten in each row (with each row made up of two groups of five), they will use the model strategically (e.g., they will move icons to make completed rows, count on, or skip-count and add on). When you notice some students beginning to use the structure of the arithmetic rack model, you can highlight their thinking by facilitating a discussion around these strategies. In this way, the model evolves into a tool *for* thinking (Van den Heuvel-Panhuizen 2003).

BIG MATH IDEAS IN THE *ATTENDANCE* ROUTINE WITH THE ARITHMETIC RACK

This version of the *Attendance* routine develops the following big math ideas:

- **Equal groups:** An essential big idea in understanding multiplication. Each row can be divided into two groups of five or one group of ten. When there are equal groups, you can use a skip-counting strategy to figure out how many.
- **Part-whole relations:** $24 = 10 + 10 + 4$ or a row of $7 = 5 + 2$.
- **Equivalence:** $5 + 5 + 5 + 5 + 4 = 10 + 10 + 4$.
- **Associative Property:** The order of adding doesn't change the whole ($10 + 10 + 4 = 10 + 4 + 10$).

For more about these and other big math ideas in early childhood, see Appendix B.

SETTING UP THE ARITHMETIC RACK *ATTENDANCE* ROUTINE

It is important for the routine materials to be physically accessible to the students. If children are doing the routines, the materials have to be at the child's level. Because wall space in many classrooms is at a premium, keeping all of your morning routines up in a meeting area often presents a challenge. One solution that teachers have found is to put the At-Home and In-School boards on foam board, bring these materials out in the morning, and put them away after the routines are done. Other teachers organize their meeting areas so that the routines they will be using consistently are permanently visible to students. In **Figure 3.1** there is a full view of how a structured model was used in a prekindergarten classroom. With this structure, the routine focused on four main questions:

1. How many boys are here today?
2. How many girls are here today?
3. How many children are present?
4. If there are _____ children present, how many milks (or napkins, plates, pencils, etc.) will we need for _____ children?

    The routine was done in a whole-group setting for the first six weeks of school. After that time, children were assigned various morning jobs that they completed in partnerships. One of these jobs was to review the attendance board after the children had come into the class each morning. This pair of students would then figure out the answers to the previous questions, record them on the chart, and then, during meeting time, report their findings to the class (Cameron, Jackson, and Zolkower 1997).

Gender and the *Attendance* Routine

Many teachers do not like classifying their children by gender. In creating this routine, we also struggled with choosing icons that reinforced gender stereotypes (boys in pants, girls in dresses or blue icons for boys, pink icons for girls). We settled on girls having a triangular base and boys having rectangular ones as a way of identifying the icons. Although we have traditionally seen gender as a straightforward way of classifying and exploring part-whole relationships in the *Attendance* routine, we acknowledge that gender is not always binary, and that students and teachers may be uncomfortable with this kind of classification. There are other ways to explore this routine that aren't about gender. For example, teachers may use other binary categories (laces/no laces; hot lunch/cold lunch). The object here is to have two parts that can be combined to make the whole (the number of children present). If there are eighteen children in the class and eight students have shoes with no laces, that means that ten students are wearing shoes that have laces.

Attendance Routine Variation 2: The Cube Stick

MATERIALS

- Unifix cubes kept in a stick to represent the total number of students in the class
- A bowl or bin with Unifix cubes

ROUTINE STRUCTURE

Using a single color of cubes, create a cube stick equal to the total number of students in the class. This cube stick stays together and is used for comparison. As children get ready to go to the meeting area, have them take one cube from the bowl and sit in a circle. When everyone is seated, have the children put their cubes in a pile in the center of the floor so everyone can see them. Ask one child to put these cubes together and figure out how many children are here today.

Using this attendance structure, children will do the following:

- Count the cubes by ones.

- Compare one cube stick to the other and use part-whole relations to figure out the number of children present. Children might say things like, "There are two kids missing today because this [the cube stick representing the children present] is two less; I can count back two cubes, twenty-four, twenty-three. There are twenty-two kids here today" or "Everyone is here today because the cube sticks match" [they have the same length].

A version of this routine can be found in the kindergarten *Investigations* curriculum (TERC 2017).

Creating a Model Using Cubes

Some teachers change the permanent cube stick to parallel the structure of the Rekenrek (arithmetic rack). Using the Rekenrek structure (alternating groups of five) helps children internalize the total quantity in relation to the five and ten structure, which is crucial for understanding base ten (place value). With the green cubes, 24 is 24 ones; with the red and white cubes, 24 can be thought of as $(2 \times 10) + 4$ or $(4 \times 5) + 4$ (or, in kid speak, 2 tens and 4 or 4 fives and 4).

BIG MATH IDEAS IN THE *ATTENDANCE* ROUTINE WITH THE CUBE STICK

This version of the *Attendance* routine develops the following big math ideas:

- **Magnitude:** One set has more; one set has less.
- **One-to-one correspondence:** The sets have the same amount.
- **Part-whole relations:** One set represents the whole (the number of kids in the class); the other set represents a part (the number of students present); the parts create the whole (when put together, the number of students present and the number of students absent is equivalent to the total number of kids).
- **Missing addend model of subtraction:** $22 + ? = 24$.
- **Removal model of subtraction:** $24 - 2 = ?$

For more about these and other big math ideas in early childhood, see Appendix B.

Although this routine uses structured materials (one cube stick represents all the children in the class; the other represents the children who are there on that particular day), there is a significant difference between the arithmetic rack attendance model and this cube stick model. The cube stick variation of this routine has some of the same big ideas as the other variations of this routine (counting, part-whole relations) but it also adds a different idea of comparing sets to think about quantity (i.e., the total number of kids in the class, which is a constant, and the number present on any given day, which can vary). However, the structure of the cube stick does not promote the development of other kinds of counting and addition strategies (e.g., counting on, grouping and adding tens) in the same way that other variations of this routine do. For example, if you compare the cube stick variation of the *Attendance* routine to the arithmetic rack *Attendance* routine, you will see that they are similar in that they both encourage work with part-whole relations. However, a key difference is that the arithmetic rack model emphasizes base ten as a structure. This base ten structure is critical for students' understanding of how our number system works and the development of critical strategies in addition and subtraction. We believe that exposure to different models through the *Attendance* routine helps students grapple with different mathematical ideas.

Attendance Routine Variation 3: The Attendance Survey

FIGURE 3.4

MATERIALS

- Survey board (**Figure 3.4**)
- Dry erase markers (children place an **X** on the survey board to show their vote)
- Attendance roster

ROUTINE STRUCTURE

When children come into class in the morning, they check off their name on an attendance roster. Then they go to the survey board and mark an **X** in the yes or no row to answer the day's survey question. Possible questions might be "Do you like pizza?," "Do you ride the bus to school?," or "Did you wear sneakers today?" Each day, after students have had a chance to respond to the survey question, two students work together to use the roster to figure out how many children are present. Using the total number of students present, they then check to see if everyone has taken the survey. They report their findings to the class.

BIG MATHEMATICAL IDEAS IN THE *ATTENDANCE* SURVEY

The routine is designed to develop some big mathematical ideas:

- **Magnitude:** One set has more; one set has less.

- **One-to-one correspondence:** If there are twenty-four students present, twenty-four students should have taken the survey.

- **Part-whole relations:** One set represents the whole (the number of kids present in the class that day); the parts are the number of students who voted yes and no. The parts create the whole (the total number of yes and no votes should be equivalent to the total number of kids present). For example, if nineteen students voted yes and five students voted no, students could be asked how they would find the sum. Here, a range of strategies are possible:

  - counting all yes and no votes from one;
  - counting on from 19 ("I used my fingers and said, 'Nineteen, 20, 21, 22, 23, 24.'");
  - reasoning with landmarks ("If 20 kids voted yes, that would mean there are 25 kids here; $20 + 5 = 25$. Because there are 19 kids and 19 is 1 less than 20, I would take away 1 from the 25. So, there are 24 kids here today.");
  - using partial sums and the associative property of addition ("I split the addends to make tens; $19 + 5 = (10 + 9) + (1 + 4) = 10 + 9 + 1 + 4 = 10 + (9 + 1) + 4 = 10 + 10 + 4$ or 24.");
  - using the structure of the model to
    - count by fives and add on the units (5, 10, 15, 20, 21, 22, 23, 24);
    - imagine the 19 as a 20, add $20 + 5$ and remove 1 from 25 to find the total number of students.

  For more about these and other big math ideas in early childhood, see Appendix B.

Learning Is Developmental

Rich contexts such as attendance offer the opportunity for young children to engage and grapple with things they do not know. For some children, the challenge may be learning to count, for others it may be discovering systems for organizing and not losing track of the count, and for others still it may be developing addition and subtraction strategies.

One of the challenges in using the *Attendance* routines is deciding which variation is just right for your students right now. For us, "just right" means that although children can *do some part of the routine*, there is an opportunity for them to make significant cognitive shifts as they grapple with a problematic situation. For example, if you have a small group of students who can count fluently and have developed cardinality, the attendance survey might be just right for them. This routine allows children to compare the sets (the children who are here and the children who took the survey) and to grapple with the challenging mathematical idea of part-whole relations. The question confronting them is "How do I figure out the missing part when I know the whole and one part?" (e.g., if there are eighteen students here today and only sixteen students took the survey, how do I figure out how many students didn't vote?). Although the mathematics in this question is challenging, students who are developmentally ready to grapple with this idea have an opportunity to make huge shifts in thinking. When choosing a routine, we always begin by considering what math students already understand and what they might be ready to take on next. We choose a routine that builds upon what they know and will also help them take on new math understandings and strategies.

Whatever the children's new learning may be, the role of the teacher is critical. Knowing when to vary the *Attendance* routine is directly connected to a teacher's understanding of the mathematical big ideas and strategies that each routine is designed to develop. With this knowledge a teacher can differentiate the learning to meet the needs of the class as a whole as well as the needs of each individual child. Knowing how to vary the *Attendance* routine and effectively facilitate conversation around the mathematical ideas is essential for developing a vibrant classroom culture in which students are able to communicate their thinking, justify their ideas, and revise their thinking. Because the soil of the *Attendance* routine is rich and nurturing, it is a place where ideas flourish.

As with all of the routines in this book, our learning as teachers grew alongside our students' learning. One teacher who used the *Attendance* routine over the course of a year reflected in this way:

> " Here's what I learned: it was *so* incredibly important for me to hold back and avoid my impulse to overscaffold the situation for learners. Whenever I felt the urge to suggest a strategy to a child, I held off for one day and reflected on the implications of doing so. For example, when a student struggled with keeping track of the children he counted during the *Attendance* routine, I wanted to jump in and help him. I held off and was surprised that other students engaged in the problem and worked with him to come up with a way to count the students. I was happy that I held off on offering help. Giving my students thinking space empowered them. It also empowered me because it gave me insights into how my teaching decisions can help students flourish. "

Developing Fluency and Flexibility with Practice Routines

" For the things we have to learn before we can do them, we learn by doing them. "

Aristotle, *The Nicomachean Ethics*

Practice Routines: A Form of Differentiation

How do you meet the developmental needs of all students? This question has perplexed, challenged, and frustrated most of us who have taught a classroom of children. And although this question cannot be answered easily, it *must* be addressed if we are to honor children's learning differences in with that ensure that each child learns and thrives (Tomlinson 2014). Addressing learning differences is with the heart of successful teaching and it is an issue with which we (Toni, Patricia, and Dani) grappled at the end of our first year implementing routines in our classrooms. To address the needs of all learners in Patricia's kindergarten classroom, we created a set of routines we call *practice routines*.

At the end of our first year using the routine *How Many Days Have We Been in School?* we assessed children's understanding of place value and the bead string as a model. (See Place-Value Interview and a transcript of one child's interview on the companion website.) Through analyzing the whole-class data from this assessment, we realized that although many children had made significant progress in understanding place value, a small group of students had not. This insight distressed us, but it also made us reflect on our practice and consider how we might change our routines to improve *all* children's learning outcomes.

As we reflected on our use of the routines in year one, we realized that there were simply too many students and too few opportunities for them to have consistent experience using the bead string model. As part of this reflection, we wondered what would happen if we gave children more consistent opportunities to practice and interact with mathematical models over the course of the school year, providing opportunities for students to develop important mathematical understandings in developmentally appropriate ways. To accomplish this goal we created a series of practice routines. This adjustment didn't mean we were excluding some children from the whole-group activities with the hundred bead string—everyone still participated in these discussions—but, rather, that we provided additional opportunities for students to work with the model. To differentiate learning we adjusted the quantity of beads (we had strings with ten, twenty, or one hundred beads) and partnered students homogeneously (Cash 2017). This structure was used to give students more opportunities to work with big ideas and strategies related to *their* development. For example, in the practice routine *Build It*, some students work with a bead string with ten or twenty beads, and others work with the hundred bead string. Over the course of the year, as students' thinking evolves, partnerships and models are changed to reflect children's development.

The good news from this work is that the combination of whole-group routines and practice routines is highly effective in developing the thinking and strategies of all children. The whole-group routines allow for rich discourse around big mathematical ideas, models, and strategies; the practice routines help level the playing field in terms of students' mathematical development, offering *all* students opportunities to explore and understand ideas in different ways and at different paces. Using these routines also helped us better assess students' mathematical thinking over time (Richardson 2012) and meet their needs during small-group instruction like guided minilessons (Fosnot and Dolk 2001).

Practice Routine: *Build It*

🎯 GOALS FOR THE ROUTINE

In the practice routine *Build It* students practice building numbers using bead strings. One of the big goals of this routine is for students to use the structure of the bead string model (the five or ten structure) as a tool for thinking (Edo and Putra 2011). Initially, students might build numbers by counting by ones. However, with meaningful practice, they start to use the five and ten structure built into the model. For example, the number fourteen might be built by pulling over ten beads on the bead string and counting up four more beads. A one-swoop action of pulling over ten beads means that students can recognize a unit of ten without counting (although they might be seeing the ten by conceptually subitizing the two groups of five) (Clements and Sarama 2009).

The three variations of this practice routine allow the teacher to differentiate, matching the practice routine variation to the students' developmental needs (e.g., if students are able to rote count to twenty, they might be assigned to play *Build It: Variation 2* to help develop number recognition and different ways to quantify numbers to twenty).

MATERIALS

- One bead string per partnership/small group; these bead strings can be ten, twenty, or one hundred beads long
- Numeral cards (see companion website)

BUILD IT: VARIATION 1

1. Numeral cards are shuffled and placed in a stack facedown.
2. Partner A turns over a numeral card.
3. Both players build the quantity shown on the card with their bead strings.
4. Players check what they've built and take turns sharing how they built that number (e.g., "It's 36. I counted by tens—10, 20, 30 and then I said, '31, 32, 33, 34, 35, 36.'" Or "I counted 10 and then I said, '11, 12, 13, 14, 15 . . . 36.'"
5. Players check that the quantities on their bead strings are the same.
6. Partner B turns over a numeral card and play continues.
7. If disagreement arises about how to build a number, partners should come to consensus about the accuracy of their quantities before continuing on with the next turn.

Supporting Students with Practice Routines

The beauty of the bead string in different quantities (a total of ten, twenty, or one hundred beads) is that students have an opportunity to work at their developmental level. Although the structure of each of the practice routines in this chapter is the same (i.e., the decks of cards may be different, but how the game is played is not), the nature of the conversation and how you interact with students will be different because the ideas they are using will be different (see **Table 4.1**: Overview of Practice Routine: *Build It* for an example of how this might sound). You might also support children's interactions by giving them some talking prompts, a clear structure for how the game is to be played and when it ends (Chapin, O'Connor, and Anderson 2009). For an example of talk prompts, see Teacher Note: Helpful Hints for *Build It*.

BUILD IT: VARIATION 2

1. Numeral cards are shuffled and placed in a stack facedown.

2. Partner A turns over a numeral card for their partner to see, then moves that number of beads to the left side on their bead string.

3. Partner B checks the bead string to see if the number of beads matches the number on the card. Partner B says, "I agree" or "I disagree" and tells his partner why ("This is nineteen. I counted one, two, three . . . nineteen.").

4. Partner A clears the bead string, hands it to their partner. Partners switch roles, and play continues in a similar fashion.

BUILD IT: VARIATION 3

1. This variation can follow the same structure as variation 1 or variation 2. However, in this version, instead of using numeral cards, students use cards that identify the units (e.g., four tens and six). See the companion website for a printable set of these cards.

2. At the end of each turn students must name the quantity (four tens and six is forty-six).

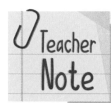
Teacher
Note

Helpful Hints for *Build It*

Organizing the Structure of the Practice Routine

Having systems for managing materials and organizing play is critical for a smooth running of this routine. A structure that we have found helpful is:

1. Partner A gets the materials; Partner B puts them away.

2. Partner A goes first and turns over a numeral card and builds the number.

3. Partner A gives the bead string to Partner B and Partner B checks to see if they agree. If Partner B agrees, they tell their partner why; if they disagree, they tell why. Both children must reach agreement for the game to continue.

4. It's Partner B's turn; repeat Steps 2 and 3 with players switching roles.

Prompts to Develop Student Talk

Although we prefer to have student conversation be organic in nature, we have found that it may be initially helpful to give young children talk prompts. These talk prompts can be placed strategically around the classroom (i.e., in the game center) or put on a "prompt sheet" that is part of the materials children get when they go to play a game. Some possible talk prompts to support children's interactions are:

1. After the quantity is built, Partner A says, "I built the number _____; do you agree?" Partner A then hands the bead string to Partner B.

2. When the bead string is handed to Partner B, the child says, "Let me check."

3. After checking, Partner B says, "I agree/disagree because . . . "

4. Once students agree on the quantity that has been built, Partner B says, "It's my turn. I'll clear the bead string for us."

5. At the end of the work session, both partners give a high five and say, "Nice game!"

TABLE 4.1   Overview of Practice Routine: *Build It*

Practice Routine	Materials	Structure of the Routine	Possible Student Strategies and Challenges
Build It: **Variation 1**	• Bead string to ten for each player • One set of numeral cards 1–10 (companion website)	1. Students play in pairs, trios, or quartets. 2. One student picks a numeral card and turns it over for everyone to see. 3. Each student builds that number and checks what they've built with their partner(s).	**Using the bead string:** • Counting all • Counting on (using five structure) • Using part-whole relations (e.g., using ten to think about eight) • Subitizing **Challenges:** • Recognizing the numerals on the cards (If cards have dots or pips on them, students may count them to figure out the number.) • Different quantities are built by kids and disagreement arises
Build It: **Variation 2**	• Bead string to twenty for each group • One set of numeral cards 1–20 (companion website)	1. Students play in pairs or trios. 2. One student picks a numeral card and turns it over for everyone to see. 3. One student builds the amount and other student/s check for accuracy.	**Using the bead string:** • Counting all • Counting on (using five and ten structure) • Using part-whole relations (e.g., using twenty to think about eighteen) • Subitizing **Challenges:** • Recognizing the numerals (students may count the dots on the card to figure out the number) • Different quantities are built by kids and disagreement arises
Build It: **Variation 3**	• Bead string to a hundred • One set of numeral cards 1–100 or one set of "Ten and . . ." cards (companion website)	Variation 3 can use the same structure as variation 1 or 2. As students become fluent with variation 3, challenge them to play with the "Ten and . . . " cards. Here, students move away from using the numeral cards to cards that identify the units (e.g., four tens and 6). After building the quantity on the card, students name the quantity (four tens and 6 is 46).	**Using the bead string:** • Counting on (using five and ten structure) • Using part-whole relations (e.g., using one hundred to think about eighty-nine) • Skip-count by tens and counting on • Using ten as a unit (this is four tens; pulling over forty beads by counting the tens, "1, 2, 3, 4, so 40") • Subitizing **Challenges:** • Recognizing the numerals • Shifting units when counting by tens (e.g., counting 10, 20, 30, 40, 50, when 32 beads are shown) • Different quantities are built by kids and disagreement arises • When using the "Ten and . . . cards", keeping track of the units may be a challenge (e.g., 4 tens and 2 might be built as 60 [the "2" is counted as two tens not two ones]). • Naming the final quantity (e.g., 8 tens and 3 should be thought of as 80 + 3 or 83).

Build It in a Kindergarten Classroom

Transcript of Conversation #1

Two students are playing *Build It: Variation 1.* The teacher watches them play for a few minutes before speaking to them.

1. **Teacher:** So interesting. I saw the two of you build eight, but you did very different things. Tori, what did Amber do?

2. **Tori:** I don't remember.

3. **Teacher:** You don't remember or you didn't see what she did?

4. **Tori:** Didn't see.

5. **Teacher:** Okay. Let's remember that it's important for both of you to watch each other and see if you understand what the other person is doing.

6. [Tori nods, yes.]

7. **Amber:** I counted the beads. [she demonstrates] One, 2, 3, 4, 5, 6, 7, 8.

8. **Teacher:** So, Amber counted the beads by ones. Amber, did you see what Tori did to build eight?

9. **Amber:** No.

10. **Tori** [clears her bead string]**:** I know this red is five. So, I can just say, "five" and not count the beads.

11. **Teacher:** Amber, did you see what Tori did?

12. **Amber:** She said, "Five."

13. **Teacher:** She said, "Five," but she did something with the beads. Did you see?

14. **Amber:** She counted.

15. **Tori:** I didn't count! I moved all the red beads over—see, this is five!

16. **Teacher:** Let me see if I can do that—you pulled over all the red beads and knew this was five.

(continues)

(continued)

Build It in a Kindergarten Classroom — Conversation #1

17. **Tori:** Yeah, just like pulling all of these beads over is ten. [points to the group of red and white beads].

18. **Teacher:** Here's my question, Tori. The numeral card says eight. What made you pull over five beads?

19. **Tori:** Five is before eight. [She demonstrates with the bead string.] All this is eight, but five—it's here. Five and this little bit more makes eight.

20. **Amber:** That's three.

21. **Teacher:** Whoa! Amber, how did you know it was three without counting?

22. **Amber:** I can see three.

23. **Teacher:** See three?

24. **Amber:** I can see four too. But I can't see eight.

25. **Teacher:** Could you show me how you "see four"? I want to be able to do that too.

26. **Amber:** This is four.

27. **Teacher:** You didn't have to count! Wow!

28. **Tori:** Hey—look! Four lives inside five. [She demonstrates by removing one bead from the group of five.] And they all live inside of eight.

29. **Amber** [laughing loudly]**:** Four lives in five. That's funny.

30. **Tori:** And they all live in eight. Here's two inside eight. Here's five inside eight. It's like eight is the house with all these beads living inside it!

31. **Teacher:** Thanks for sharing. Numbers live inside each other—do you think you two could share your strategies with the class later? Amber, you can share how you just know four without counting; and Tori, you can share how numbers live inside each other.

Analysis of Teaching Moves

Watching and waiting before conferring helps the teacher understand students' ideas and prepare her opening question. Notice that the teacher started with Amber's strategy to highlight an important community norm: everyone's ideas matter and the job of each learner is to pay attention to what other learners are doing and saying.

In line 8, the teacher names the strategy Amber used. In lines 11–18, the teacher highlights that Tori's strategy was different, and slows down to explore it. This is an effective teacher move on three levels. First, the teacher is establishing norms for working together—we watch each other and share how we build. Second, she models curiosity about a strategy she "doesn't understand." Third, she models that sometimes learners need to try things they don't fully understand yet ("Let me see if I can do that"). Here, she's modeling how slowing down to explore an idea can be an important tool to help make sense of a situation.

In line 18, the teacher goes below the surface of Tori's strategy to explore a big idea in math (i.e., hierarchical inclusion, that five is nested inside of eight). To use the five structure to build eight, a child must understand that five is inside eight or how five can be used to make eight. These are the first steps in understanding that in an equation like $5 + ? = 8$, 8 is the set that can be broken down into two parts, 5 and 3.

In lines 21–27, the teacher celebrates Amber's subitizing strategy, and this move helps Amber share her ideas. This subtle move not only builds a student's confidence, but it lets both children know that everyone's ideas are important and can be celebrated.

In line 31, the teacher repeats Tori's words, "numbers live inside each other," and asks both girls to share their strategies in the whole-group discussion. This is an important move that can support a wide range of learners. For some children, who are counting all, using subitizing as a strategy will be a big shift. For others, the idea that you can use the colors on the bead string to name a set (e.g., pull over five because it's a group distinguished by the color) will support the development of counting on strategies (e.g., "This is five; I can count on six, seven, eight"). And for most children in the class, working with part-whole relations is a big idea they will continue to work with in early childhood and beyond.

(continues)

(continued)

Transcript of Conversation #2

Two pairs of students are sitting at a table playing *Build It: Variation 3*. The teacher watches both pairs play and decides to have one group share a strategy they're using with the other group.

1. **Teacher:** I've been watching all of you play. I want to say it's wonderful how you're listening to each other and checking what's been built. I saw Juliana do something really interesting—I was fascinated by your conversation with Jed—and I'd like you to show Paul and Mica what you did when you built eighty-six. [Paul and Mica put down their bead string and watch.] And, Juliana, since you built eighty-six differently from how Jed did it, I'm going to ask Jed to share what he did and then what you did.

2. **Jed** [clears the bead string]: I built 86 by counting [pulls over the beads as he counts] 10, 20, 30, 40, 50, 60, 70, 80 and then I did [moves over single beads as he counts] 81, 82, 83, 84, 85, 86.

3. **Teacher:** Paul and Mica, that's how I saw you building numbers too.

4. **Mica:** Yup. We count tens; then we count ones. Same way.

5. **Juliana** [excitedly]: I did something different! [She reaches for the bead string.]

6. **Teacher:** I can see how excited you are, Juliana—it's wonderful to make new discoveries. But I'm wondering if you could allow Jed to build it—to see if he understands.

7. **Juliana** [reluctantly and unhappily]: Okay.

8. **Teacher:** I know it isn't easy to let someone share your idea—thanks for being such a good partner and letting Jed share your idea.

9. **Jed:** She said, "One hundred!" [As Jed starts to demonstrate the rest of Juliana's strategy, he is stopped by his teacher.]

10. **Teacher:** Before Jed does that, Paul and Mica, I'm wondering how pulling over one hundred beads might help you build eighty-six. [Both Paul and Mica remain silent for a moment.]

11. **Paul:** I don't know.

Build It in a Kindergarten Classroom — Conversation #2

12. **Teacher:** What about you, Mica?

13. [Mica shrugs.]

14. **Teacher:** It's really challenging to think about this, isn't it?
[Both children nod, yes.]

15. **Jed:** I didn't get it either when Juliana did it the first time. I was like, 'One hundred—you went too far.' But then she showed me: this is one hundred. This is ninety [removes ten beads]. This is eighty [removes another ten beads]. Then she said, "Eighty-one, 82, 83, 84, 85, 86."

16. **Teacher:** Would you like a chance to try this strategy, Paul and Mica?

17. **Paul:** Yes.

18. **Mica:** I don't know.

19. **Paul:** This is one hundred, right, Mica? [Mica remains silent.] When I use all the beads, it's one hundred. Look, 10, 20, 30, 40, 50, 60, 70, 80, 90, 100. All the beads over, it's one hundred. [Paul counts the units of ten.] One, 2, 3, 4, 5, 6, 7, 8, 9, 10. Ten tens is one hundred.

20. **Juliana:** And eighty is eight tens. Ten tens is one hundred; eight tens is eighty. Do you agree [to Mica and Paul]?

21. **Paul:** Ten, 20, 30, 40, 50, 60, 70, 80.

22. **Juliana:** That's eight tens. [points to each group of ten and counts]
One, 2, 3, 4, 5, 6, 7, 8.

23. **Jed:** Yup. So, you can go to one hundred and go back two tens. Here's ten tens; here's eight tens.

24. **Teacher:** This feels like a lot. I'm feeling a little shaky here—Mica, how are you doing?

25. **Mica** [barely audible]**:** Shaky too.

26. **Teacher:** It's good to know how you're feeling. Using a one hundred to build eighty-six is a new strategy and new strategies are challenging to use. It's important to know how we're feeling and to say, 'Hey, I'm not ready to do that yet.' It's probably the most important thing you can do as a learner. Ask people to slow down. Say what you know, what you don't know. You may not be ready to

(continues)

(continued)

Build It in a Kindergarten Classroom — Conversation #2

use a different strategy and that's okay. Paul, if you feel ready to use it, go for it. Here's a challenge for you [to Juliana and Jed] . . . With what numbers does it make sense to use this strategy—to go to one hundred?

27. Jed: You can use it with any number.

28. Juliana: No. I think only some.

29. Teacher: You don't have to answer that question yet—think about it. It sounds like you disagree—and that's great. Now you have to convince each other. Jed says you can use it with any number. Juliana says you would use it with only some numbers. Keep playing and talking; I'd love to hear what you have to say about it later on.

Analysis of Teaching Moves

The teacher watches the children play before speaking and then names the positive in what they're doing. This reinforces the community norms she is trying to build. These norms encourage students to listen to one another, watch each other closely, and share strategies. The teacher also helps students manage their impulsivity (in lines 6–8) when she asks Juliana to let her partner demonstrate. This kind of sharing and turn taking isn't easy for five-year-olds! The teacher emphasizes what she values when she thanks Juliana for giving her partner a turn; she also helps her understand that turn taking isn't an easy thing to do when you're excited by a new strategy you've just discovered.

In lines 1–9, the teacher highlights what strategy the children have in common (counting by tens and counting on) before introducing the new strategy. Here, she invites Jed, Juliana's partner, to demonstrate it, but she stops when he pulls over one hundred beads. This slowing down of the strategy is an important move for helping the other children make sense of what's happening. A new idea has come into the community and it may be one that some children need a lot of time to explore and understand.

Notice the time she gives Paul and Mica to understand the strategy and how, at the end, she lets the children know that it's okay not to know. She underscores what Mica is feeling by saying that learning is a challenge and it's up to every learner to decide when they're ready to try something new.

She ends her conference by challenging Juliana and Jed to think about when a strategy like this would be useful. Everyone leaves the conference with a challenge; this is the essence of differentiation!

Practice Routine: *Sequencing Numbers*

🎯 GOALS FOR THE ROUTINE

The goal of *Sequencing Numbers* is for students to become fluent in recognizing and sequencing numbers from least to greatest and greatest to least (Nunes and Bryant 1996).

Students begin by arranging the cards in sequence from 1 to 10 or 1 to 20 with complete decks (e.g., all the numbers from 1 to 10 or 1 to 20). As students become fluent with organizing cards within the counting sequence, they are given partial decks. As they sequence these cards, students gain insights into patterns in the number system as they think about magnitude (is fourteen greater than ten?). For example, if a student pair has turned over the cards fourteen, eight, three, and ten, they have to think about which number is the least and which number is the greatest. Initially, they may have to figure this out by counting from one to know how to sequence the cards ("One, 2, 3, 4, 5, 6, 7, 8 . . . oh, 3 comes before 8 when I count"). However, with practice, students begin to think about the digits and the meaning of the digits in a number (e.g., 10 is less than 14; 14 has a 10 and 4 more in it; 8 is greater than 3 because 8 is made up of 5 and 3). When students become fluent in working with numeral decks 1–20, they can also practice sequencing numbers 1–100.

MATERIALS

- Numeral cards (one set per partnership); range of decks are 1–10; 1–20; 1–100 (see companion website)
- Numeral cards 1–20 and 11–100 can be partial decks (The numeral sequence is not complete; for example, a deck of 1–20 could have only odd or even numbers or a mixture of both odd and even numbers; a deck of cards to 100 might have all the multiples of 10 or might have two cards per decade—for example, 8 and 9, 18 and 19, 28 and 29—or the multiples of 10 plus or minus 1—for example, 9, 10, 11, 19, 20, 21.)

ROUTINE STRUCTURE

1. Students work in pairs to arrange the numeral cards in sequence.
2. After sequencing the cards, children read the cards from least to greatest or from greatest to least.

Helpful Hints for *Sequencing Numbers*

Organizing the Practice Routine

We do not recommend sequencing complete decks from one to one hundred because there are simply too many cards for children to manage and sequencing so many cards provides space challenges. There are, however, different ways to organize the cards (e.g., one deck has only the cards from twenty to forty or one deck has only the multiples of tens and numbers that are one more and one less than each ten), so that children are practicing counting and sequencing numbers greater than twenty. To help organize these materials, we recommend storing these sets of cards in labeled bags (e.g., "20–50," "tens + 1 – 1"). It is also helpful to have the card sets copied on different colors of cardstock so that when children put them back in the bags, the sets do not get mixed up. Although the materials for all of these practice routines take time to prepare, we've found that all the work you put into organizing things at the beginning is enormously beneficial to managing the practice routines over time.

Prompts to Develop Student Talk

It is often challenging at first for students to explain their thinking when they are doing these routines. Teaching students prompts for speaking and responding to what is said can be helpful initially. We try to not give students too many prompts—we know we've done this when we hear them trying to remember *what they're "supposed" to say* rather than engaging naturally with their peers. Remember, the goal with prompts is for students to eventually move away from them and have more authentic kinds of conversations. **Table 4.2** lists some prompts and words that we have found helpful. **Table 4.3** provides an overview of the *Sequencing Number* practice routine.

TABLE 4.2 *Sequencing Numbers:* **Possible Talk Prompts**

Child Does	Child Says	Partner Response	Helpful Words
Puts cards in order	"I put my cards in order. [Child reads the numbers in the sequence and gives a rationale for his choice.] I did this because . . . [e.g., 3 is less than 14 and 14 is less than 18]."	"I agree with you because . . . " or "I disagree with you because . . . [14 is bigger than 18]."	• *Put in order* • *Less/smaller than* • *Greater than/bigger than* • *More than* • *Agree/disagree because* • *After/before* • *Let's check* • *Your turn*
	"I'm not sure I agree. Let's check. We can use the number line to see if 14 is less than 18."		
Gets the number line	"See—it says, 14, 15, 16, 17, 18. Eighteen is after 14."	"I agree. fourteen is before 18; 18 is after 14 when we count on the number line."	
	"It's your turn."		

TABLE 4.3 **Overview of Practice Routine:** *Sequencing Numbers*

Practice Routine	Materials	Structure of the Routine	Possible Student Strategies and Challenges
Sequencing Numbers: Variation 1	• Numeral cards 1–10 (one complete set per partnership) (see companion website)	**Variations 1–3** 1. Students work in pairs to sequence the numeral cards. 2. After sequencing, partners read the cards from least to greatest or from greatest to least.	**To sequence numbers students may:** • Count from one (have to figure out where a number goes by going back to one and counting up to the number on the card) • Recognize and use patterns (one-digit numbers; two-digit numbers; numbers less than twenty; numbers greater than twenty, etc.) • Use specific relationships (e.g., 19 is 1 less than 20; 22 is in between 20 and 30, but closer to 20) • Focus on before and after ("What number comes right after six?") • Figure out which numbers come between other numbers ("What number comes between six and eight?") **Challenges** • Recognizing the numerals • Counting forward and counting backward by ones in the correct number sequence Language development: *before/after; least to greatest; greatest to least*

(continues)

TABLE 4.3   Overview of Practice Routine: *Sequencing Numbers*

(continued)

Practice Routine	Materials	Structure of the Routine	Possible Student Strategies and Challenges
Sequencing Numbers: Variation 2	• Numeral cards 1–20 (one complete set per partnership) (see companion website)	**Variations 1–3** **1.** Students work in pairs to sequence the numeral cards. **2.** After sequencing, partners read the cards from least to greatest or from greatest to least.	**To sequence numbers students may:** • Focus on before and after ("What number comes after sixteen?") • Figure out which numbers come between other numbers ("What number comes between sixteen and eighteen?") **Challenges** • Recognizing numerals, especially the teen numbers • Knowing the number words for teen numbers (e.g., *15* might be said as "fifty") • Counting forward and counting backward by ones in the correct number sequence • Language development: *before/after; least to greatest; greatest to least*
Sequencing Numbers: Variation 3	• Numeral cards 1–20 or 1–100 (one partial set per partnership) (see companion website)		**To sequence numbers students may:** • Count to place a number ("Twenty-two is before 26 . . . if you count it goes 20, 21, 22, 23, 24, 25, 26.") • Organize cards using the structure of the number system (For example, students might say, "All the numbers with a 2 in the tens place are less than numbers with a 3 in the tens place" or "A number with a 9 in the ones place always comes before a ten—19, 20, 29, 30, 39, 40.") • Focus on which numbers are between other numbers ("Here's 16 and 28; where would 22 go? Is it more than 16? Less than 28?") **Challenges** • Recognizing numerals • Knowing the number words for teen numbers and multiples of ten (e.g., *15* might be said as "fifty") • Language development: *before/after; least to greatest; greatest to least*

Practice Routine: *Guess My Number*

🎯 GOALS FOR THE ROUTINE

Students practice recognizing and sequencing numbers in order to guess the missing number.

MATERIALS

Numeral cards (one set per partnership); range of decks are 1–10; 1–20; 1–100
(see companion website)

Note: Numeral cards 1–10 and 1–20 are full decks. The deck of cards for 1–100 is a partial one made up of the multiples of 10 or the multiples of 10 plus or minus 1 (e.g., 9, 10, 11, 19, 20, 21, 29, 30, 31) or other counting sequences (e.g., the cards 20–40).

ROUTINE STRUCTURE

1. Partner A picks and hides a card from the numeral deck.

2. Partner B sequences the cards in the deck to guess the missing number.

3. Partner B guesses which card is missing and states her reasoning ("I think [number] is missing because . . . "). Partner A shows the card she was hiding.

TABLE 4.4   **Overview of Practice Routine:** *Guess My Number*

Practice Routine	Materials	Structure of the Routine	Possible Student Strategies and Challenges
Guess My Number: **Variation 1**	• Numeral cards 1–10 (one complete set per partnership)	1. Partner A picks and hides a card from the numeral deck.	**To sequence numbers students may:** • Organize the cards in the counting sequence (forward or backward), counting aloud as the numerals are placed in order
Guess My Number: **Variation 2**	• Numeral cards 1–20 (one complete set per partnership)	2. Partner B sequences the cards in the deck to guess the missing number.	• Use before and after as a strategy (For example, pick up the card with 6 on it and say, "This is 6, then comes 7; 7 is after 6. Here is 5; 5 is before 6.") • Start with one number and build up (For example, "Here's five. Then comes 6, 7, 8, 9, 10.")
Guess My Number: **Variation 3**	• Numeral cards (one partial set per partnership) (see companion website) See Teacher Note: Helpful Hints for *Sequencing Numbers* for more information on choosing numbers for partial sets.	3. Partner B "guesses" which card is missing and states their reasoning ("I think [number] is missing because . . . "). Partner A shows the card they were hiding.	**Challenges** • Recognizing numerals • Knowing the number words for teen numbers and multiples of ten (e.g., *15* might be said as"fifty") • Counting forward and counting backward by ones in the correct number sequence

Practice Routine: *Less and More*

🎯 GOALS FOR THE ROUTINE

Students practice sequencing numbers using a card to think about what number is one less or one more than (or before/after) the number on the card. Students also practice writing numbers.

MATERIALS

- Numeral cards (one set per partnership); range of decks are 1–10; 1–20; 1–100 (see companion website)
- Laminated number line in ones 0–10; 0–20
- Laminated hundreds chart (for the cards 0–100)
- Game boards: (1) *Less and More* and (2) *One Less and One More*
- Dry erase markers

ROUTINE STRUCTURE

1. The number line or hundreds chart is turned facedown.

2. Partner A takes a card from the numeral deck and puts it on the center space of the game board.

3. Partner B names the number that is less than the number on the card (e.g., "Six comes before 7; 6 is less than 7"); Partner A names the number that is more than the number on the card (e.g., "Eight comes after 7; 8 is more than 7").

4. Children write their numbers on the game board after saying them.

5. Once they've built their sequence, children turn over the number line or hundreds chart and check their sequencing.

6. Children turn the number line or hundreds chart facedown again and erase their game board.

7. Play continues in this fashion.

TABLE 4.5 Overview of Practice Routines: *Less and More* and *One Less, One More*

Practice Routine	Materials	Structure of the Routine	Possible Student Strategies and Challenges
Less and More: Variation 1	• Numeral cards 1–9 (one set per partnership) (see companion website) • Number line 0–10 • *Less and More* game board	1. The number line or hundreds chart is turned facedown. 2. Partner A takes a card from the numeral deck and puts it on the middle space of the game board. 3. Partner B takes a card and decides where the card goes on the game board. (If 10 is in the middle space and the child turns over a 16, the 16 card is placed on the "More" space and the child says, "Sixteen is more than 10.")	**Student strategies** • Sequencing numbers • Focus on more and less ("What number is more than six?") **Challenges** • Recognizing numerals • Knowing the number words (For example, in Variation 2 teen numbers may be challenging; *15* might be said as "fifty." In Variation 3, *82* might be read as "twenty-eight.") • Language: *more/less*
Less and More: Variation 2	• Numeral cards 1–19 (one set per partnership) (see companion website) • *Less and More* game board • Number line 0–20	4. Partner A turns over a card and decides where the card goes on the game board. (If the 14 card is turned over, the card is placed on top of the 16 card on the "More" space and the child says, "Fourteen is more than 10." 5. Play continues until there are no more cards. 6. Children can use the number line or hundreds chart to check their work, as necessary.	
Less and More: Variation 3	• Numeral cards 1–100 (one set per partnership) • *Less and More* game board • Hundreds chart		
One Less, One More	• Numeral cards (choose a set appropriate for the group of students) • *One Less, One More* game board • Hundreds chart or number line	1. The number line or hundreds chart is turned facedown. 2. Partner A takes a card from the numeral deck and puts it on the middle space of the game board. 3. Partner B names the number that is 1 less than the number on the card (e.g., "Six is 1 less than 7"); Partner A names the number that is 1 more than the number on the card (e.g., "Eight is 1 more than 7"). 4. Children write their numbers on the game board after saying them. 5. Children turn the number line or hundreds chart faceup and check their sequencing.	**To figure out what number is one less and one more students may:** • Use counting strategies from one to figure out what number is one more or one less • Add or subtract one • Use patterns in the number system

FIGURE 4.1 Full-size versions of game boards are available on the companion website.

Less and More Game Board

Less		More
	Number Card	

One Less and One More Game Board

One Less		One More
	Number Card	

FIGURE 4.2 Examples of the *Match the Model* routine

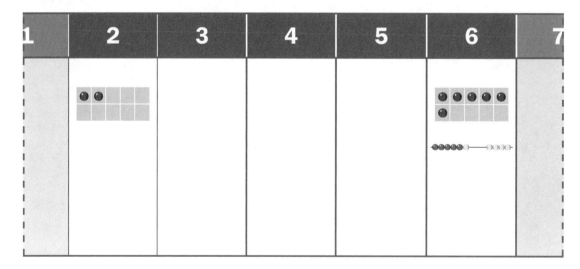

Practice Routine: *Match the Model*

🎯 GOALS FOR THE ROUTINE

This routine is designed to help students use the five and ten structure of the ten-frame and bead string models to recognize quantities from one to ten. For example, when turning over a card that shows six, a child might use the structure to say, "I know this is five [either pointing to the top row of the ten-frame or the group of red beads on the bead string] and one more makes six." This routine supports *counting on* as a strategy. It also develops part-whole relations (e.g., children use the visual structure to think of a quantity deconstructed into parts—7 is 5 and 2 more; 9 is 1 less than 10). The development of this kind of part-whole understanding is essential for the development of addition and subtraction strategies.

MATERIALS

- Ten-frame or bead string cards (1–10) (see companion website)
- Game board with numbers 1–10 written on it (see companion website)

ROUTINE STRUCTURE

1. The cards are turned facedown.
2. Partner A takes a card from the deck and figures out how many there are in total.
3. Partner B checks and says, "I agree" or "I disagree" and says why.
4. Once both partners agree on the quantity, the card is placed on the numeral game board.
5. Play continues in this fashion. The routine ends when the game board is filled with ten-frame or bead string cards covering in the space below each numeral.

   See Figure 4.2 for an example and Table 4.6 for an overview of the *Match the Model* routine.

TABLE 4.6   Overview of Practice Routine: *Match the Model*

Practice Routine	Materials	Routine Structure	Possible Student Strategies and Challenges
Match the Model: Ten-Frame	• Ten-frame cards (1–10) (see companion website) • Game board with numerals 1–10 (see companion website)	**1.** The ten-frame cards (or the bead string cards) are turned facedown. **2.** On her turn, each player turns over one ten-frame card (or one bead string card) and figures out how many dots (or beads) are on the card. **3.** The player then matches the card to the number on the numeral game board and places the card in the appropriate box. **4.** Play ends when the board is filled with ten-frame or bead string cards.	**Student strategies** • Counting all the dots/beads • Using the five structure and counting on • Using the ten structure and counting back • Subitizing • Part-whole relations (e.g., "I know this is nine because nine is one less than ten") • Basic facts (e.g., "I know three and three is six")
Match the Model: Bead String	• Bead string cards (0–10) (see companion website) • Bead string match-up board		**Challenges** • Counting the dots/beads accurately • Matching the quantity to the numeral • Recognizing the numeral (e.g., some students may confuse 9 with 6, for example)

Formative Assessment During Practice Routines

When assessing students as they engage with the practice routines, it is important to remember that they are *playing* with mathematical ideas related to our number system. Since they are playing, we want them to describe their thinking in the moment, while they're at play. To keep the spirit of play alive, it's important for *you* to be playful when you confer with students (Pinchover 2017). Although we may not often think of assessment as a playful time, the way we interact with students as they engage with the math has the power to encourage them to be curious and try out new ideas and strategies. For an example of what this might sound like, see the two conferences on pages 71–76.

Balancing Whole-Class and Small-Group Routines

In figuring out how to differentiate instruction for all of the students in our classes, we found that using a balance of whole-class and small-group routines was effective. Although whole-class routines introduce students to big math ideas and provide opportunities for whole-class discourse, the practice routines allowed partners and small groups to focus on fluency and flexibility with these ideas. How? First, practice routines enabled us to differentiate without diminishing the content (e.g., with *Build It,* everyone was building quantities; however, some children needed to build quantities to twenty while others worked with quantities greater than twenty). Second, children were matched with activities in which they could gain fluency; their fluency built their feelings of competence because nothing fuels success like success! Third, although every child progressed within a developmental sequence, they did so at their own pace. Fourth, expectations for language development and communication were consistent across the routines. Whether they were building fifteen or eighty-nine, whether they counted by ones or tens, the expectation was that students would explain to their partner how they knew the quantity was what they said it was.

Students made progress in their understanding and strategies because we provided experiences that allowed them to grow into mathematical ideas at their own pace. We also found that small-group instruction and partner work are critical companions to the work we did with whole-class routines. Some children who were hesitant to share ideas during the whole-class routines thrived in small groups and partnerships because they felt safe to speak, to try on new language and ideas, and to be supported by their peers. We also designed some additional practice routines for the whole-group setting. To read more about these routines, called *Counting Games*, see the companion website.

One of the great joys in teaching is the feeling that you, as a teacher, not only understand how your students think and where they are developmentally, but can provide experiences that meet them where they are so that they can ultimately succeed. We found this joy with the practice routines.

The *Is It Fair?* Routine

> " I am just learning to notice the different colors of the stars, and already begin to have a new enjoyment. "
>
> **Maria Mitchell, scientist**

Jimmy [four years old, looking at a photograph]: Is that your son?

Teacher: Yes, this is my son, Mark. He's in high school.

Jimmy: He's big! [Jimmy is puzzled for a moment.] Are you older?

Teacher [laughing]: Yes, I was about twenty-four years old when he was born.

Jimmy: If you're older, how come he's bigger?

One important way children develop an understanding of their world is by noticing, naming, comparing, and evaluating. As they notice and evaluate what's happening around them, they often use comparative language to describe their observations. Their noticings are often rooted in the language of opposites: tall and short; big and little; more and less. These comparisons are how children perceive similarities and differences, make sense of relationships, and communicate what they notice to others (Sophian 2008).

As young children play and interact throughout the day, we often hear them say things like, "You have more blocks than I do!" "I have the most pretzels!" "I'm taller than you." "I'm the fastest." All of these observations have implied opposites (e.g., I can have more only if you have less, or I can be fast only if someone else is slow) and come naturally to most children. Because children easily make comparisons, we may forget that the mathematical ideas underlying some of these comparisons are quite complex.

To think about this underlying complexity, let's return to the conversation in which Jimmy is examining and comparing his teacher and her son. Using the photo, Jimmy is making three comparisons: (1) a family relationship (mother and son); (2) age (the mother is older than her son); and (3) height (the son is taller than the mother). Although some of what he notices fits into his worldview and how he has generalized specific relationships (there is a mother and her son; the mother is always older than the child), one observation puzzles him: how can a younger person be taller than an older one? What Jimmy believes or *knows* is based on his own lived experience and what he intuits from it—he is aging and simultaneously growing; therefore, logically, when you age, you grow taller. His hypothesis is that a mother, who is older, should be taller than her child. Because he has conflated age and height, he experiences disequilibrium when what he *thinks* is true comes into contact with the reality shown in the picture (Pine, Messer, and St. John 2001).

Comparison: A Big Idea in Early Mathematics

Although not all comparison words that young children learn are related to mathematics (e.g., *soft* and *loud*, *noisy* and *quiet*, *black* and *white*, etc.), many are. It is important to remember that even when children make comparisons using mathematical language, it doesn't necessarily mean they fully understand the mathematical ideas underlying their words. For example, height and age are complex measurement ideas, abstractions, that are beyond the mathematical comprehension of many young students. Although age is a more abstract idea than height (e.g., you can perceive how tall someone is and compare that height to another height; with age, the comparison is more abstract—what actually shows how old someone is?), understanding linear measurement is still challenging, even though one can "see" different lengths and compare them. For more about how children compare mathematically in early childhood, see the companion website.

Using Comparative Contexts in Early Childhood: *Is It Fair?*

Although comparisons come naturally to children as they play and make sense of their world, not all kinds of comparison are equal mathematically. We have found that comparisons of quantities are easier for students to understand than comparisons of more abstract ideas like length, height, or age. Although there are many opportunities for students to compare quantities during the school day, we have found that snack time in particular offers rich mathematical possibilities for comparing quantities. In comparing their shares, students inevitably ask, "Who has more?" and "Is it fair?" As children explore various snack situations and try to make things fair, they begin to engage with some powerful mathematical ideas.

We found using snack in a routine to be a gold mine for the development of not only students' mathematical reasoning but also their ability to create justifiable arguments around their solutions. The routine we developed, *Is It Fair?*, emerged from our observations and explorations of children's thinking within this context. To help you think about the mathematical potential in this type of routine, let's peek into a kindergarten classroom.

> **Max** [looking at his snack and comparing his pretzels with those of the other children at his table]: Sabrina has more than me!
>
> **Teacher:** Are you sure?
>
> [Max nods yes.]
>
> **Teacher:** How do you know?
>
> **Max:** It looks like more. [He counts.] I only have 1, 2, 3, 4, 5.
>
> **Sabrina:** I have eight. I counted them already. He has less; I have more.
>
> **Max:** It's not fair—she has more!
>
> **Teacher:** Hmm. This seems like a real problem. I'm wondering how we might make it fair, Max.
>
> **Max:** You can give me more pretzels.
>
> **Teacher:** Okay. I could give you more pretzels, but how many pretzels should I give you to make it fair?
>
> **Max:** A lot.
>
> **Teacher:** How many is that—a lot?
>
> **Max** [thinks for a moment]: Eight.

Teacher: Sabrina, do you agree? Should I give Max eight more pretzels?

Sabrina: Yeah [pauses]. No, wait. If you give him eight, he'll have mine and his. That's not fair [points to her eight pretzels and Max's five]. He'll have all of this! [Sabrina indicates how their pretzels will be combined into a new set.]

Teacher: This seems like a real problem. Sabrina, you have more than Max and Max has less than you. Max wants to have the *same* number of pretzels as you. I want to make it fair, but I'm not sure how I can do that. Maybe this is something we can bring to our class meeting today. We can ask everyone to think about this very interesting problem about how to make snack fair. For now, Max, I'll take these five pretzels and put them in a baggie for later. Here are the eight pretzels you wanted because you thought this would make snack fair.

Max [counts as the pretzels are put on his napkin and happily says]**:** I have eight too. [to Sabrina] Now we're the same.

What's the Math?

As children struggle to make things fair, they have an opportunity to grapple with some central ideas in early mathematics:

1. **Magnitude (more or less) and one-to-one correspondence (sets are equal; there's the same amount in each).** You have more pretzels than I do. But we both have five.

2. **Cardinality (the total quantity or the last number counted names the set).** How many pretzels do I have? How many do you have?

3. **Comparison of sets (a way to know if it's fair).** Do we have the same amount? Who has more? Who has less? How much more or less?

4. **Equivalence (there are equal amounts in each set).** If you have more, how do we make it fair (make our sets equivalent)?

5. **Part-whole relations (a set [the whole] can be decomposed into different parts).** I have five pretzels; you have eight. You have three more than I do because my five is part of your eight; I need three more to have the same amount as you.

What Makes Equalizing Our Snack Challenging?

Equalizing sets is challenging for young children. In equalizing tasks students are not just asked to quantify sets, but also to compare quantities in relationship to each other. The easiest type of comparison is by eyeballing—if it looks like she has more, she has more. However, for Max to answer the question of how many more pretzels he would need to make the amount fair, he has to quantify both sets and then compare the amounts to each other. Although he can easily count each set and compare the quantities (eight pretzels is more than five pretzels), making the sets equal is still difficult for him (Nunes and Bryant 2009).

Children have intuitive strategies for approaching these kinds of problems. Some children compare by lining up two sets of objects in a direct comparison. They point to the "more" part of one set as what they need (e.g., when comparing eight and five, they use one-to-one correspondence to line up the items and point to the three more as what they need for the situation to be fair). This is a strategy and solution that do not necessarily involve quantifying each set in relation to the other.

Equalizing sets is challenging for young children because it requires an understanding that there is a relationship *between the sets*. This relationship can't be quantified—it's an abstraction (Kamii 2000). To equalize the sets, Max has to understand that he has fewer pretzels than Sabrina (see **Figure 5.1**). The difference between their sets is three pretzels; this three refers to the missing quantity between the sets that will make them equivalent. This is a complex idea because "a difference is not a quantity; it is a relation" (Nunes and Bryant 2009, 5). Although we can count things to quantify, we cannot count the relationship between things. This isn't to say we can't nudge children to understand that there is a match between the quantities they have in common (both children have five pretzels) and that the three extra pretzels that Sabrina has needs a match in Max's for both sets to be equivalent. However, as we saw with Max and Sabrina's snack conversation, this idea is not immediately apparent to the children because it requires a major shift in how they think.

Teacher Note

Using Snack as a Context

Although our experiences using snack as a context for the *Is It Fair?* routine have been positive and powerful, we recognize some teachers may want to use a context that does not involve food. Whatever you decide, remember that the structure of the routine can be used with other kinds of items (cubes, blocks, etc.). Whatever item you choose, it is important that children feel invested in making the situation fair. Student investment is generated by rich routines; it is the essential ingredient that sustains learning and enables you to use that routine over time in different ways and for different mathematical purposes (Boaler 1993).

FIGURE 5.1

Comparing Quantities (A Possible Student Strategy for Comparing)

SABRINA'S EIGHT PRETZELS

MAX'S FIVE PRETZELS

Understanding That Those Quantities Are Related in Specific Ways

SABRINA'S EIGHT PRETZELS

MAX'S FIVE PRETZELS

There are two ways to think about the three pretzels to the right of the dotted line in **Figure 5.1**: (1) how many more pretzels Sabrina has than Max or (2) how many more pretzels Max needs to have the same quantity as Sabrina. Here are some of the big ideas in early number that students have to understand to answer these questions:

$8 = 5 + 3$ **Part-whole relations, hierarchical inclusion, equivalence, and one-to-one correspondence**

Sabrina's set of eight can be deconstructed as part of the comparison. In the snack context, Sabrina's eight pretzels can be compared in a one-to-one correspondence to Max's pretzels. They have five pretzels in common. Sabrina's "more" is the part Max does not have (where the one-to-one correspondence ends). They have equivalent amounts of five; they also have unequal amounts (Sabrina has three and Max has zero). Sabrina has three more pretzels than Max, or Max has three fewer pretzels than Sabrina.

$5 + ? = 8$ **Missing part [addend]; part-whole relations**

In the snack context, Max has five pretzels and the question mark in the equation represents how many more he needs to have the same amount as Sabrina. Five is what he has; eight is what he wants, and there's some more he needs to get to have that eight. The five pretzels he has is the only tangible quantity; the other pretzels are imaginary. This means he has to hold on to two things in his head (what he has and what he wants) while he figures out the difference between these quantities. That's pretty abstract thinking for a five-year-old!

Notice how much reflection is involved in understanding the relation between sets in this problem—these types of mental actions are very challenging in early childhood (Kamii 2000).

Launching the *Is It Fair?* Routine

There are different ways to launch the *Is It Fair?* routine. We have found that using real snack food grounds the routine in children's "lived experiences" (Bransford, Brown, and Cocking 2000). The snack becomes the manipulative for children to make comparisons. The *Is It Fair?* routine sequence in **Figure 5.2** is one way to develop children's thinking and move from concrete to abstract.

FIGURE 5.2 – PART I

Is It Fair? Routine Sequence: Part I

Same Amount of Snack for Each Child

Time Frame	Structure of the Routine	Role of Teacher	What's the Math?
Two to three days	• Give out snack to all children. • Be sure you give an *equal* amount to each child. • Change the quantity on each day (e.g., Day 1 everyone gets six pretzels; Day 2 everyone gets three pretzels; Day 3 everyone gets five pretzels).	**1. Setting Norms** Establish community norms for snack. In addition to rules around cleanliness (e.g., we wash our hands before eating), there are other important norms to establish. Two important ones for this routine are: • We wait to eat our snack until everyone is served. • We don't touch other children's food. **2. Language Development** This routine provides many opportunities for students to develop comparative language (e.g., *less, more, the same*). Students may also be comparing quantities (e.g., "I have six. You have the same as me"). **3. At the end of each snack time, take time to discuss some (or all) of the following questions:** • What did you notice about snack today? • What questions do you have? **4. Assessment** As you observe students during snack time, *Listen for:* • Development of comparison language • Conversations around how many, more, etc. • Who is comparing quantities over time (e.g., "Yesterday we had six. Today we only have four!") *Watch for:* • Counting and comparing strategies • Organization (e.g., do students put their snack in a row before they count?)	**Counting strategies** • One-to-one tagging • Accurate rote counting (knows the sequence of numbers and can say them orally, but may not sync what's said with each object counted) • Accurate counting (voice touch and tag in sync) **Big ideas** • Magnitude (more/less) • Organization (e.g., putting pretzels into a line to make counting easier and to keep track)

Notes for Facilitation:

Before you move to Part II of this routine, be sure to note which children are beginning to quantify and compare the snack. Putting at least one child at each table who is already interested in talking about these topics will jump-start the conversation about unequal quantities.

For example, a child who is very observant of what everyone is getting will quickly realize when they do not get the same amount as other children. To make this happen, you might want to give two pretzels to that child and six to a neighbor. As children notice what's happening, as they compare, they are bound to bring up that the snack situation is not "fair."

FIGURE 5.2 – PART II

Is It Fair? Sequence of Lessons: Part II

Unequal Amount of Snack for Each Child

Time Frame	Structure of the Activity	Role of Teacher	What's the Math?
At least three days	• Give out snack to all children. • Be sure you give an *unequal* amount to each child. • Be sure to put at least one child whom you have observed counting and comparing at each table for this part of the routine.	**1. Setting Expectations for Community** Continue developing community norms for snack as discussed in Part I of this routine. If children are upset when they realize they got a different amount of snack than their tablemates, reassure them that you will be having a discussion about whether this is fair and figuring out a way to fix this problem as a class. **2. Language Development** This routine provides many opportunities for students to develop comparative language (e.g., *less, more, the same*). Students may also be comparing quantities (e.g., "I have four." "You have six." "You have more than me.") **3. At the end of each snack time, take time to discuss some (or all) of the following questions:** • What did you notice about snack today? • What questions do you have? • What strategies did we come up with for making snack fair? **4. Assessment** As you observe students during snack time: *Listen for:* • Development of comparison language • Students who compare quantities over time (e.g., "Yesterday I had six. Today I only have four!") • Students who remark about the "fairness" of the situation • Students who are unhappy or frustrated by the quantity of snack they are given • Students who are comparing quantities (e.g., comparing their snack to other children's snack, using words like *fair, not fair, more, less*) • Students who are engaging their peers in conversations around more, less, fair, not fair, etc. Engage children in conversation to see how they are thinking about (1) the meaning of *fair* and (2) how to make things *fair*. The conversations you have one-on-one or in small groups about these ideas can be brought to the whole group for consideration.	**Counting strategies** • One-to-one tagging • Accurate rote counting (knows the sequence of numbers and can say them orally, but may not sync what's said with each object counted) • Accurate counting (voice touch and tag in sync) **Big ideas** • Magnitude (more/less) • Organization (e.g., putting pretzels into a line to make counting easier) • Equivalence (How do we make snack "fair" or create equal sets?) • Hierarchical inclusion (e.g., "She has six; I have four. She has more. When you count you say, four, five, six"). • One-to-one correspondence (e.g., He has four and I have four too. We have the same.")

(continues)

(continued)

FIGURE 5.2 – PART II

Is It Fair? Sequence of Lessons: Part II

Unequal Amount of Snack for Each Child

Time Frame	Structure of the Activity	Role of Teacher	What's the Math?
		Watch and Listen for: • Counting and comparing strategies • Organization (e.g., if they put their snack in a row before they count)	

Notes for Facilitation:

As you facilitate conversations around children's strategies for quantifying and comparing snacks, be sure to note children whose voices and ideas may tend to dominate the group discussion. We want to make sure to call on children with *different* kinds of strategies and understandings. Allow students to grapple with how to make snack fair and what it means to be fair. Remember, this routine is about allowing students to develop strategies and build on each other's understandings, not "fixing" incorrect or inefficient thinking. It can be tempting to try to "fix" things as an adult who has lots of experience with this kind of problem, but that ultimately robs children of agency and struggle. Remember, *productive struggle is the doorway to new and deeper understandings.*

If you are worried that some children in your class might be living with food insecurity at home or may be particularly sensitive to the idea of not getting "enough," remember that in this routine everyone is getting snack and that

the issue of fairness is resolved as a community pretty quickly. This is not to say children always find ways to equalize the amounts, but that their natural ingenuity allows them to come up with solutions that are equitable. For example, in one classroom, a child offered this solution: "Everyone should get five pretzels; that's not too much and it's not too little!" Alternatively, some teachers have used a context for this routine that is not food related. Keeping the question "Is it fair?," some teachers have made the routine context about sharing classroom materials (e.g., books, blocks, cubes, pencils). One teacher created a story about Billie and Bobby that was about snack in *their* classroom; the story was used to help students visualize the problem that Billie and Bobby were having and think about how they might make *that* situation fair. Another teacher collected stickers that her local Trader Joe's grocery store gives out for free. She used stickers as the context for her *Is It Fair?* routine.

FIGURE 5.2 – PART III

Is It Fair? Routine Sequence: Part III

Is It Fair? with Images

Time Frame	Structure of the Routine	Role of Teacher	What's the Math?
Ongoing. You will use this version of the routine going forward; This routine can be used once a week or more frequently.	• Use different images of snack situations, rather than actual snack food. • The question "Fair? Or not fair?" will frame the discussions you have with children. • If you use this routine as a quick image, "Fair? Or not fair?" are questions that may be asked but not resolved with each image. For example, you might discuss each image, or you might show several images in a row and then return to one for further discussion about fair or not fair. • Some possible images for this routine are included in **Figures 5.3** to **5.6.** slide decks of these images as well as additional images for classroom use are available on the companion website.	**1. Setting Expectations for Community** Establish community norms around how to listen to and respect the ideas of each learner. This is especially important in this part of the routine because children who know differently than their peers (e.g., "If I have 4 and you have 6 and we want to make it fair, I can't give you 6 because then you would have 4 *and* 6") may at first tend to dominate conversations or be inconsiderate of what they know is a "wrong" answer. *Considerations for norms around listening and talking:* 1. What does it mean to *listen?* • We repeat what someone is saying. • If we don't understand, we ask a question. • We can say what we *do* understand and what we *don't* understand. (This moves children away from saying "I don't get it" to saying things like, "I understand [this part of what he said], but I don't understand [that part of what he said]"). 2. How do we disagree/agree with someone? • We can check that we understand someone's idea by asking a question ("Are you saying . . ." "Do you mean . . ."). • We can say, "I disagree with _____ because . . . " or "I agree with _____ because . . . " • We can disagree with an idea, but still respect the person sharing the idea. **2. Language Development** Within this routine, there are opportunities for developing students' language and ability to ask questions. Because the emphasis is on sharing ideas, children will be developing their tool kit for communication. Key here is developing: • clarity of expression; • clear mathematical language; • the ability to paraphrase what has been said, not just saying what they want to share (or what they wanted the other person to say).	**Counting strategies** • One-to-one tagging • Accurate rote counting (knows the sequence of numbers and can say them orally, but may not sync what's said with each object counted) • Accurate counting (voice touch and tag in sync) **Big ideas** • Magnitude (more/less) • Organization (e.g., putting pretzels into a line to make counting easier) • Equivalence (how to make snack "fair" or create equal sets) • Hierarchical inclusion (e.g., "She has six; I have four. She has more. When you count you say, four, five, six.") • One-to-one correspondence (e.g., "He has four and I have four too. We have the same.")

(continues)

FIGURE 5.2 – PART III

Is It Fair? Routine Sequence: Part III

Is It Fair? with Images

(continued)

Time Frame	Structure of the Routine	Role of Teacher	What's the Math?

	There are many ways to compare two sets. Here are some ways to compare that we explore in the routine:
	
	• same item; different quantity;
	
	• same item; same quantity, different use of space;
	
	• different item; same quantity;
	
	• different item; different quantity.
		3. After the routine take time to discuss some (or all) of the following questions:
		• "What did you notice about the snack situation we thought about today?"
		• "What questions do you have?"
		
		4. Assessment
		As you observe,
		Listen for:
		• Specific social and mathematical language
		• Conversations around
		  - how many, more, fewer etc.
		  - the "fairness" of the situation
		  - proof-like analysis and generalizations about *all* snack situations (e.g., "If I have three, and you have any amount more than three, it's not fair")
		
		Watch and Listen for:
		• Counting and comparing strategies
		• Proof (how they prove who has more and justify their solutions)
		• If children engage their peers in conversations around more, less, fair, not fair, etc.	

Types of Images for the *Is It Fair?* Routine

Although there are many ways to explore this routine, we have found a few types of images to be helpful in developing students' ideas, reasoning, and mathematical language within this snack context. These four types of image are explained in **Figures 5.3**, **5.4**, **5.5**, and **5.6** and included in the *Is It Fair?* slide deck on the companion website.

FIGURE 5.3

Image Type #1

Everything is the same (same snack type, same quantity of snack, same visual arrangement of snack)

Example of Possible Image

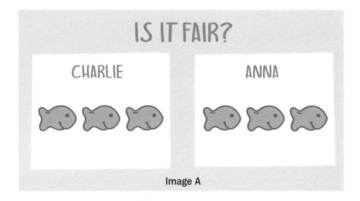

Image A

When to Use This Image

This is an early image to use with students. It can be used in two different ways:

1. as an image where you ask students to describe what they see;

2. as a quick image (flashed for just a few seconds). You can say to students, "I'll show you Charlie and Anna's snacks quickly and then you will decide whether their snacks are fair or not fair."

What's the Math?

Big Ideas

- Subitizing
- One-to-one correspondence
- Equivalence

Language Development

- *Same*   • *Horizontal*
- *Equal*   • *Middle [of the napkin]*
- *Row*

FIGURE 5.4

Image Type #2

Same snack type and quantity; different visual arrangement of snack

Example of Possible Image

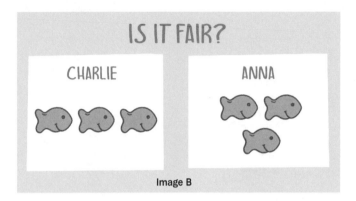

Image B

When to Use this Image

Although this image may appear straightforward to adults, expect that some children will not recognize Anna's snack as three. They may describe it as two and one. This image can generate rich mathematical conversations if there is disagreement about the total quantity and whether or not the situation is fair. Questions like, "How is Charlie's snack the same as Anna's snack? How are they different?" are useful. Also, challenge students who say the snacks are the same to prove *how* they are the same. As part of this "proof," you may ask students to rearrange one of the images to show that even though the snacks look different, the situation is fair because everyone gets the same amount of snack (Schultz-Ferrell, Hammond, and Robles 2007). This rearranging can be done by projecting the image on an interactive whiteboard or simply by having physical images of the snacks on magnets that students can move around on a magnetic board.

What's the Math?

Big Ideas

- Subitizing
- One-to-one correspondence
- Equivalence
- Part-whole relations
  (two and one is three)

Language Development

- *Same/different*
- *More/less/equal*
- *Row(s)*
- *Horizontal*
- *Middle* (of the napkin)

FIGURE 5.5

Image Type #3

Same snack type and quantity; various sizes and shapes of snack are shown

Example of Possible Image

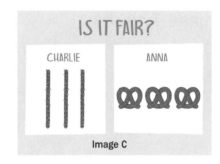

Image C

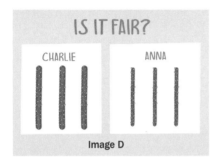

Image D

Image E

When to Use this Image

There may be disagreements with this type of image about whether or not the situation is fair. Use these disagreements to help students understand that sometimes *multiple* ways of thinking can be correct. This kind of image is an example of an ambiguous situation that leaves it up to the interpreter to justify his position (Danielson 2016).

Image C can be used to generate big ideas around quantity (e.g., that 3 = 3) and that the amount of snack can be a separate attribute from the type of snack itself. In this set of images, one can consider the attributes as (1) quantity; (2) type of pretzel; (3) density (thick/thin) and, (4) height (tall/short). Also, expect for students to grapple with issues of fairness related to size (is the tall pretzel "more" than the short one?).

A common argument for Image D is that you get more snack with the *fatter* pretzel and, therefore, although both children are getting three pretzels, Charlie is eating a little more pretzel than Anna (Greenes, Ginsburg, and Balfanz 2014).

Image E is designed for students to develop spatial awareness and specific language related to how the image is organized. For example, children might describe what they see by saying, "There are three pretzels at the top of the napkin" or by using their hands to indicate that although Charlie has three pretzels in one row, Anna has three rows of pretzels. As new language arises from children, keep track of the words that help them describe what they are seeing.

What's the Math?

Big Ideas

- Subitizing
- One-to-one correspondence
- Equivalence

Language Development

- *Same/different*
- *More/less/equal*
- *Row(s)*
- *Horizontal/vertical*
- *Middle (of the napkin)*
- *Tall/short*
- *Thick/thin*
- *Big/little*

FIGURE 5.6

Image Type #4

Same type of snack, but different quantity

Example of Possible Image

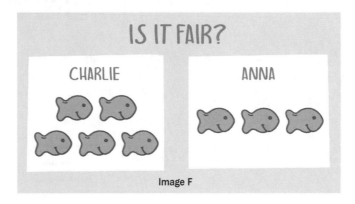

Image F

When to Use this Image

This image type is designed for students to compare sets and to discuss fair or not fair and ways to make the situation fair. Notice how the goldfish in Image F are organized in a way that draws attention to the three in a row (e.g., how one set has the other set *inside* of it). This is a big idea related to hierarchical inclusion and part-whole relations. It is okay if children are not immediately able to resolve the dilemma of how to make the situation fair. With experience and opportunities to talk to one another, they will develop more strategies for equalizing sets (Mercer 2008).

What's the Math?

Big Ideas

- Subitizing
- One-to-one correspondence
- Equivalence
- Hierarchical inclusion (three is inside of five)
- Part-whole relations (two and three equals five)
- Missing part ($3 + ? = 5$)

Language Development

- *Same/different*
- *More/less/equal*
- *Row(s)*
- *Horizontal*
- *Middle* (of the napkin)
- *Inside*

How Many Ways Can We Compare?

Ms. K.'s kindergarten classroom has been thinking about the question "Is it fair?" once a week for several months. As new strategies for comparing quantities are offered by children during the *Is It Fair?* routine, Ms. K. keeps track of them on a chart titled "Our Strategies for Knowing if It's Fair." Together with the class, she builds the chart in **Figure 5.7** with a visual above a description of the student's strategy (Guarino et al. 2019). Over time, this chart became a powerful reference, and students refer to each other's strategies by name ("I used Ebony's strategy and found what was the same—that's how I knew it wasn't fair").

FIGURE 5.7   Strategy Chart for *Is It Fair?*

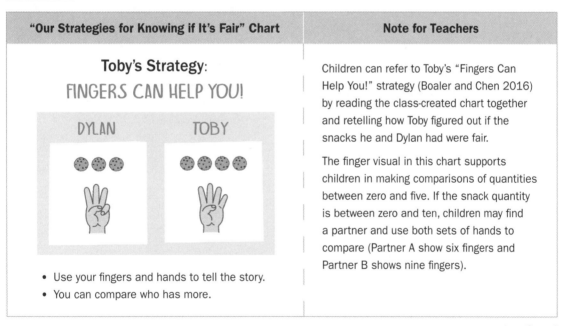

"Our Strategies for Knowing if It's Fair" Chart	Note for Teachers
Toby's Strategy: FINGERS CAN HELP YOU! DYLAN TOBY • Use your fingers and hands to tell the story. • You can compare who has more.	Children can refer to Toby's "Fingers Can Help You!" strategy (Boaler and Chen 2016) by reading the class-created chart together and retelling how Toby figured out if the snacks he and Dylan had were fair. The finger visual in this chart supports children in making comparisons of quantities between zero and five. If the snack quantity is between zero and ten, children may find a partner and use both sets of hands to compare (Partner A show six fingers and Partner B shows nine fingers).

(continues)

FIGURE 5.7	Strategy Chart for *Is It Fair?*

(continued)

"Our Strategies for Knowing if It's Fair" Chart	Note for Teachers
### Lucia's Strategy: LONGER DOESN'T MEAN MORE! 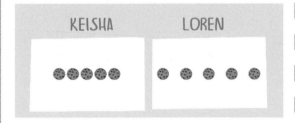 • Longer doesn't mean more. • When I'm not sure, I can count to check.	Although Lucia's comparison strategy (count to check if the amounts are the same) is an important one, it may not emerge the first few times you do this routine. That is just fine, as it is important to let strategies emerge from the children. Lucia's strategy caused a lot of debate when it first arose, and some children, even at the end of the discussion, were not convinced that Loren didn't have more snack! Our job as teachers in these kinds of situations is to remain neutral so that children can have ownership of the debate. Given the opportunity, children will try to convince each other of their arguments and this kind of mathematical debate is central to learning.
### Ebony's Strategy: FIND WHAT'S THE SAME 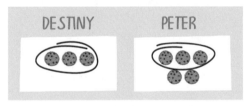 • You don't have to count all the cookies. • Look for what's the same. • If someone has extra, then they have more.	Children often look for what's equivalent between sets; subitizing these quantities helps them think about who has more (Sarama and Clements 2009). Sometimes, to prove the logic of their thinking, children come up to the image and hide "the extra part" to prove that if what remains is the same, then "the extra" they are hiding means someone has more (Sophian 2008).
### Alex's Strategy: LOOKS LIKE THE DOTS ON DICE 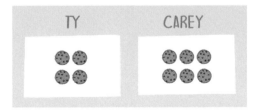 • Dice can help us! • I know how four looks. I know how six looks. I don't have to count to know which is more.	Children often transfer what they know from games and other math experiences to the *Is It Fair?* routine. They may use dot patterns on dominoes and dice and also refer to geometric shapes (e.g., "There were three cookies that look like a triangle") as tools to compare quantities.

What Does It Mean to Be Fair?

The idea that something is fair or not fair is complex for a number of reasons. First, although we are using the context of fair/not fair around snack and equal amounts, fair is a social construct that children need to understand both in and out of school. It's important to be open to what children are bringing to the table from their lives outside of school and from their own perspectives about how the world works (Gonzalez, Moll, and Amanti 2005). For example, we once had a child argue that the teachers should get more snack because "they're bigger and bigger people should get more than little people." Although we acknowledged this child's understanding of fair as valid and thoughtful, we explained that during our snack time routine we would be thinking about fair as equal.

As part of this routine, we have also had children make comments that we did not pursue in whole-class discussion. Once a child said (as a way to explain why it was okay for her to have more): "My dad says life isn't fair, so get used to it!" Another child explained why she shouldn't worry if he had more: "We have a rule in my house. You get what you get, and you don't complain." Although these comments led to interesting discussions among the children, we took a listening role in these discussions. We did this to honor that there are many ways of understanding the idea of fairness and to help children understand that different people will have different ways to determine if something is "fair." Although we did not jump in to resolve the issue, we did encourage children to discuss these ideas around fairness in a whole-class meeting. Here, the children came to consensus about what they thought it meant for snack to be fair: "Fair means we all have the same amount of food."

Assessing Students' Thinking Using the *Is It Fair?* Context

Throughout the year, we formatively assess students' comparison strategies using different written tasks (Van den Heuvel-Panhuizen 1996). These tasks allow us to see how students are taking on different comparison strategies that have been shared during the routine (see **Figure 5.8**). We use information from these assessments to help us plan for further work within the *Is It Fair?* routine and other comparison tasks. In one first-grade class, we assessed students' thinking in December. **Figures 5.9**, **5.10**, and **5.11** show three examples of children's unique strategies.

FIGURE 5.8 A written *Is It Fair?* task allows teachers to consider next steps for work with comparative problems.

> The teacher was giving out snack.
> Anna got 15 goldfish and Will got 9 goldfish.
> Will said, "Hey, that's not fair!" The teacher said, "How can we make it fair?"
> How would *you* make it fair? Show your thinking in the space below.

FIGURE 5.9 Solution One

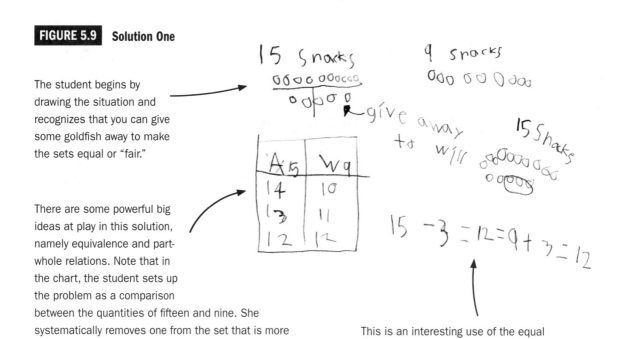

The student begins by drawing the situation and recognizes that you can give some goldfish away to make the sets equal or "fair."

There are some powerful big ideas at play in this solution, namely equivalence and part-whole relations. Note that in the chart, the student sets up the problem as a comparison between the quantities of fifteen and nine. She systematically removes one from the set that is more and gives it to the set that is less. This compensation strategy maintains the set or the whole (twenty-four goldfish in all). She stops removing snacks when both children have the same amount, twelve.

This is an interesting use of the equal sign that illustrates a number of ideas at play: $12 = 12$; $15 - 3 = 12$ *and* $9 + 3 = 12$; therefore $15 - 3 = 9 + 3$.

FIGURE 5.10 Solution Two

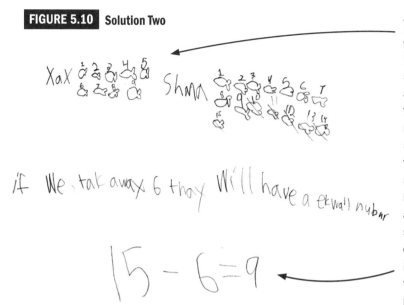

if We tak away 6 thay Will have a ekwall nubur

15 − 6 = 9

This comparative strategy examines how many more fifteen is than nine. The child draws the situation, numbers each goldfish, and crosses off the "extras" in the set of fifteen. Although he represents his strategy using the subtraction equation, 15 − 6 = 9, the essence of what he is doing seems more rooted in comparison: he finds nine in the larger sets and crosses off the extra. The six in his equation is found by counting what's been crossed off. He does, however, write "if we tak away 6 thay will have a ekwall nubur," which might indicate that he recognized the crossing off as a kind of subtraction, a removal.

FIGURE 5.11 Solution Three

Jason   pnal
13   7 +
         |
      7+6=13

You can give 6 for pual equall 13. now pual have 13 and me and pual both have 13.

This piece of work is from a different first-grade classroom (note the numbers and students' names are different). The strategy used here is to think about the relationship between thirteen and seven and find a way to equalize the sets so that the child with seven has thirteen. The mathematical structure behind this strategy is that of a missing addend subtraction problem: 7 + ? = 13. When the student finds the missing part, she writes an equation, 7 + 6 = 13, to show how many more goldfish need to be given to the child with seven to make the situation fair (e.g., give six more goldfish to the child with seven goldfish. Then both children have thirteen goldfish).

Using Contexts in Routines to Develop Powerful Math Thinking

In both kindergarten and first-grade classrooms, the *Is It Fair?* routine helped children develop an understanding of challenging math ideas. As these students moved on to the next grade, some of our second-grade teacher colleagues asked us what we were doing; they wondered why these particular groups of children were more easily able to solve complex subtraction comparison problems than others have been in the past.

We think the answer to our colleagues' questions lies in the types of routines we use consistently in our classrooms. For example, the *Is It Fair?* routine offers children ongoing experiences with complex mathematical ideas like part-whole relations, and problem structures like missing addend. We believe that the power of this routine is directly related to the fact that it's rooted in a real context—one that makes sense to children and is highly relevant to their everyday lives. Students not only have multiple opportunities to reason about unfair situations during the *Is It Fair?* routine but are also totally invested in solving the problems they encounter. They have time to share and justify their thinking and listen to that of their peers. In the messiness of agreeing and disagreeing, children shift their perspective and develop more sophisticated ways of reasoning (e.g., "If you have more, I just need to count up [or add on] from my quantity to yours to make the situation fair").

Finally, we don't believe that these routines in and of themselves will transform children's reasoning. The key ingredient is *you*! If you believe in the capacity of all children to learn, and give them the space and encouragement to do so, they will not only meet, but more than likely exceed, your expectations. Remember, there are no weeds in this garden—just many different plants. They all need water to grow.

Using Quick Images to Develop Reasoning, Strategies, and Big Ideas

Quick Images: A Peek into a Powerful Routine

Ms. Clark's prekindergarten class is buzzing with excitement because it is time for math, and they have been working on a quick image routine called *Is It the Same? Is It Different?* Children are seated in the meeting area, waiting for Ms. Clark to show them the first image.

FIGURE 6.1

Image A remains the same
throughout the routine.

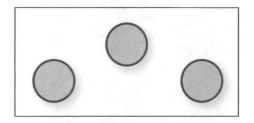

Image B is one of several images that
students compare to Image A using
Is It the Same? Is It Different?

Ms. Clark: Yesterday we did a new quick image routine called *Is It the Same? Is It Different?* Does anyone remember that routine? [Hands shoot up into the air.] Wow! That's a lot of hands and some of you—I can tell by your faces—are *very* excited.

Charlie [calling out]**:** I love this game!

Ms. Clark: That's wonderful, Charlie, but let's try to remember to use our hand signals. Remember what happens when we all talk at once [puts her hands over her ears] . . . Yikes! It's hard to hear. Okay, so what were the rules from yesterday's quick image game? Analise?

[Analise doesn't speak, but indicates a thumbs up and a thumbs down.]

Ms. Clark [imitating the hand signals]**:** So, we used hand signals. We used a thumbs up and a thumbs down. When were we using those signals, Analise? Do you remember?

Analise: You put up a picture—it was four dots that looked like this [indicates the dots with her two hands], two here and two here. Then we looked at other pictures and had to show if we thought they were the same [indicates a thumb up] or different [thumb down].

Ms. Clark: Thanks, Analise, that's exactly what we did. I'm going to show you an image—a picture—and I'll put it up here on the board for everyone to see. [She puts up Image A from **Figure 6.1**.] And then I'm going to show you some other images and you're going to use your thumbs to speak for you—quiet mouth, talking thumb. [Students giggle.] If your thumbs are up, you're saying, "The images are the same." If your thumbs are down, you're saying, "The images are different." What does this signal mean [shows the class a thumbs down]? Everyone?

Class: It's different.

Ms. Clark: And this [shows thumbs up]?

Class: It's the same.

Ms. Clark: Great. Here we go. Here's the first image. Looks like everybody is ready. I'm going to count to three and when I say three, everyone's eyes should be up here to see the image. One, two—everyone should be looking up here now—three [shows Image B from **Figure 6.1** for five seconds]. Is it the same or is it different than our image that's on the board? Thumbs up if you think it's the same; thumbs down if you think it's different. [Some thumbs are up, and some thumbs are down.] That's interesting. Some thumbs are up; some thumbs are down. We don't agree. And you know how I love a good debate! Who's ready to share their thinking about why they think the image is the same or why they think it's different? Jackson, why don't you start us off?

In this short vignette, Ms. Clark is using a type of quick image designed to help students develop their abilities to reason mathematically and communicate their thinking. Before reading the rest of the classroom dialogue, think about why there is disagreement about to whether the two images in **Figure 6.1** are the same or different.

Jackson: It was the same.

Tyler: Different.

Jackson [shouting]**:** Nuh-uh! Same!!

[Tyler shakes his head no vehemently.]

Ms. Clark: It's okay to disagree. We may be seeing things differently. We may have different ideas. So, Tyler and Jackson disagree, but we don't know why. Can you explain to each other—and to everyone else—why the images are the same or why they're different?

Jackson: It was three. [Jackson counts the dots in Image A.] Look—One, two, three. Three dots. Then it was [indicating the dot placement in Image B with his finger in the air] one, two, and three was up here [indicates the dot was above the other two].

Tyler: It looked different. One is like this [indicates a row with his hand motion] and the other looks this [indicates a triangular shape with his two hands].

Jackson: That doesn't matter; it's still three dots.

Ms. Clark: That's so interesting. [She draws a T-chart on the board and labels columns *Same* and *Different*.] So, these two images are the *same* if we think about how many dots are here [puts both images together in the column labeled *Same*]. As Jackson said, there are three dots. And he counted to prove it to us. He said, [points to the dots] one, two, three for this image. And then, one, two, three for this other image. But Tyler said the images were *different* if we think about how they look—they do look very different! [She moves the two images under the column labeled *Different*.] In this image, the dots look like this [mimics making a row with her hand as Tyler did].

Kyra: The other one looks like a triangle!

Ms. Clark: It looks like a triangle. And in this other image, the dots look like this [gestures that the dots are in a row the way Tyler did]. There's a math word that describes things that look like this [indicates the row]. Does anyone know what it is? Dawn?

Dawn: Line.

Ms. Clark: Line would work to describe these dots. We could also say they're in a row. [She points to the two images as she speaks.] Here the dots are in the shape of a triangle and here the dots are in a row. Tyler, could you please remind me to add the word *row* to our math word wall? Thanks! Okay. Who's ready to try another one? [Students wave their hands in the air.] Okay. Here we go! Ready? One, two, three . . .

Ms. Clark spends just about ten minutes with this routine, and yet in this short amount of time she is helping her students develop some important learning behaviors (Costa and Kallick 2000). Mathematically, students are learning about big ideas such as set (that a quantity of things has a number name), *conservation* (that three is three no matter how it *looks*) and *one-to-one correspondence* and *equivalence* (that if two sets have a one-to-one match, they are equivalent to each other). Socially, students are learning to be members of a community of discourse where everyone's ideas matter (Sherin 2002). Since value is placed on students explaining their thinking and being able to justify their answers, they are challenged to find ways to convince each other (Russell et al. 2017).

This routine illustrates a shift away from math experiences in which there is only one right answer (Dweck 2006). As teachers, we can encourage this shift by choosing images for which an argument can be made that the images are both the *same* and *different*. This use of quick images nudges children to deal with ambiguity and consider alternate points of view. Even more importantly, learning how to listen to and appreciate different ideas is a lesson that extends far beyond this routine.

A Deeper Dive into Quick Images

What are quick images? As the name implies, quick images are images that are shown quickly! These images can be made up of a variety of things (e.g., blocks, dots, shapes) that are structured in different ways. When an image is shown quickly—for about three to five seconds—the way in which the objects are organized impacts what the eye is drawn to and how the image is visualized. In early number, quick images can be designed to help students visualize numbers as quantities and form mental representations of sets. As such, quick images can be used as instructional tools for moving students beyond counting by ones. Using some images repeatedly (e.g., an image that looks like the 5 on a die) can develop a kind of automaticity with quantity (think of sight words in literacy). Combining these sight images can help students develop early addition strategies and fluency with the basic combinations (Wheatley 1999). Additionally, when used strategically, quick images can help students construct big ideas like part-whole relations and equivalence, which are critical to understanding addition and subtraction. Quick images can also be used to develop students' academic language. In this chapter, we will share how quick images can be used to help students think mathematically and develop their ability to communicate their thinking with clarity. For further reading on quick images, see Clements and Sarama (2009), Shumway (2011), and Wheatley (1999).

Using the Structure of Quick Image Routines to Impact Thinking

In our work with quick images, we have explored the role-specific structured models (e.g., Rekenrek, ten-frame, bead string, dice, and domino) play in developing students' mathematical reasoning and communication (Gravemeijer 1999). A routine's structure—this includes the questions, prompts, and the types of models used—greatly influences children's perceptions and strategies. For example, if we change the prompts or questions we use, the same image can provoke different kinds of mathematical reasoning! As we noticed the impact of teacher language and types of models on student thinking, we began to think carefully about which quick images we used and why and how we were using them. To help us plan, we created categories or types of quick image routines.

Although there are no hard-and-fast rules about when to use each type of quick image, there are ways to think about why some routine types might be thought of as precursors to others. For this reason, we recommend that teachers become familiar with the different ways quick images might be used (e.g., some for language development; others to develop a big mathematical idea like part-whole relations) and to consider how (and why) to use these from a developmental point of view. See **Appendix E** "Assessment: If This, Then That" for more about deciding which kind of quick image to use and how to use them.

What's the Math: Why Quick Images in Early Childhood?

One of the first challenges children encounter in early number is that numbers are not just names but have *numerosity* (a quantity or set connected to each number name). **Figure 6.2** illustrates the difference between number *as name* and number *as quantity*:

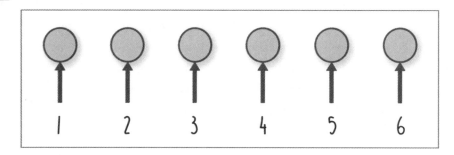

A child counts six counters but does not yet connect the number name with the quantity. When asked to point to the six counters, he points to the sixth counter in the row.

To assess whether students have developed the idea of set or cardinality, you might ask the child to take six counters from a pile of counters. After the child has made this set, ask them, "Can you show me three?" If a child has not yet developed the idea of set (*cardinality*), they may point to the third counter. For children who think this way, three represents a name, not a quantity (e.g., the third counter is the "three"). To understand that three is a quantity of objects, the child must develop a way of thinking about how each object is embedded within that set. This big idea is called *hierarchical inclusion* (Fosnot and Dolk 2001). When asked the question "Can you show me three?" a child who has developed an understanding of hierarchical inclusion might show you any three counters that are part of the set of six, pulling three counters from the left, the right, or the middle. Without an understanding of set, children often think it matters where you start counting (e.g., you have to count from left to right—you cannot start on the right because that counter is "six") or that you might get a different count if you started at the end of the row (e.g., with the sixth counter). The visual in **Figure 6.2**, when juxtaposed with the visual in **Figure 6.3**, highlights the difference in understanding between number as name and number as quantity:

FIGURE 6.3

This visual is an example of hierarchical inclusion (Kamii 2000).

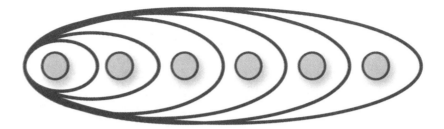

Notice that each object in **Figure 6.3** is discrete, but also a part of the set. Two objects are part of the set of six; three objects are part of the set of six as well. As a quantity, one is nested inside two; five as a quantity is nested inside six. This big idea, hierarchical inclusion, is necessary for developing additive thinking and part-whole relations, both critical understandings for addition and subtraction (six is the total of the set, but it can be deconstructed into parts that, when put together, equal six—one and five, two and four, three and three, etc.).

Because developing cardinality (in counting, the last number spoken names a set or quantity) is critical in children's early number development, routines that promote this kind of reasoning are essential. Quick image routines help students subitize quantity (see a group of objects as a set) and are perfect tools for developing cardinality. For more information on the big ideas of early number discussed in this section, see **Appendix B**.

Teacher Note

The Big Ideas of Mathematics

Big ideas are written about in different ways by different mathematics educators and researchers. We prefer the Fosnot and Dolk (2001) metaphor of a "landscape of learning," which is made up of big ideas (structures), strategies (how one solves something), and models (tools for manipulating or thinking with). For different viewpoints about big ideas, see the writing of Randall Charles (2005) or Marian Small (2010).

"Is It Six?": Developing Children's Cardinality

José is a prekindergarten student. His teacher is using a one-on-one interview to assess his counting strategies and understanding of cardinality. She gives José six counters.

1. **Teacher:** How many counters do you have, José?
2. **José** [puts the counters in a row and counts slowly from left to right, pointing to each counter]: One, 2, 3, 4, 5, 6.
3. **Teacher:** How many is that?
4. **José** [recounts the organized counters from left to right]: One, 2, 3, 4, 5, 6.
5. **Teacher:** What if I moved this counter over here? [The teacher moves one counter slightly away from the group and places it so that it is not in the row of counters.] How many counters are on the table now?
6. **José** [counts the remaining counters left in the row]: One, 2, 3, 4, 5.
7. **Teacher:** What about this counter?
8. **Josée:** That's one.
9. **Teacher:** How about if I moved it back here [puts it back in the row with the other counters]? How many counters on the table now?
10. **José** [counts the remaining counters left in the row]: One, 2, 3, 4, 5, 6. Six!
11. **Teacher:** Final question, José. [She points to the sixth counter in the row José has made.] If I started counting the counters from here, how many counters would there be?
12. **José:** You can't do that.
13. **Teacher:** I can't? Why not?
14. **José:** 'Cause that's six. [José grows impatient.] Look! [He counts from left to right again.] One, 2, 3, 4, 5, 6. [He points to the counter farthest to the right.] That's six—not one!
15. **Teacher:** Could you show me the six?
16. **José** [picks up the sixth counter]: This is six.

"Is It Six?": Developing Children's Cardinality

17. **Teacher** [holds up the same counter]: Is this six counters or one counter?

18. [José is silent, starts to fidget.]

19. **Teacher** [puts the counter back down]: When I put this counter here [puts it back in the row], it's how many?

20. **José** [counts again]: One, two—six. This is six [points to the sixth counter].

José's thinking illustrates just how complex it is for young children to develop the idea of set (cardinality). Often students struggle to identify the meaning of six—is it a label ("I am six" or "I take the 6 train")? Is it a name I say when I count? (Van den Heuvel-Panhuizen 2008). To understand six as a set is one of the big shifts in children's thinking in early number.

Structuring Quick Images

The beauty of using quick images is that they can be structured in certain ways that can nudge children to develop a variety of strategies for answering the question "How many?"

STRUCTURING THE IMAGE: THE "HOW MANY?" QUESTION

Let's start by comparing the two images in **Figure 6.4** and thinking about how each image is structured. Before reading on, consider these questions:

1. How would you figure out how many dots are in each image?

2. What would you expect children to do and say to answer the "How many?" question?

FIGURE 6.4 **Structuring quick images**

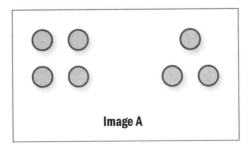

Image A

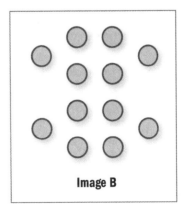

Image B

You probably noticed that Image B is more visually complex than Image A. Although both quick images require subitizing to answer the question "How many dots?," the groupings in Image A are much easier to see because of the clear space between them and because the group of four dots is represented in the familiar dice pattern (Sarama and Clements 2000). With Image B, you have to mentally create the space between the parts of the image. You have to deconstruct the whole image into parts. In Image B you might use the symmetry that is built into the visual (whether it's dividing the image in half vertically or horizontally) to construct the whole from the parts. You might also focus on the two fours that make up the inner part of the image and add on the four dots that are to the right and left of these fours. No matter how you see Image B, a lot more work needs to be done to reconstruct this image and answer the question "How many dots?"

Because Image A presents groups of dots that are easier to subitize than Image B, a *How Many Dots?* routine with Image A might focus on using the structure of the image to help children develop cardinality and move beyond counting by ones. When Image A is shown quickly, children have to make a mental snapshot of the image's organization. They can use the parts of Image A to figure out the total number of dots. Although a child may be able to subitize the groups of three and four dots, they may still need to use a counting or adding strategy to figure out the total number of dots.

One possible strategy for figuring out the total number of dots in Image A is counting all ("I saw 4 and 3 and I said, 'One, 2, 3, 4, 5, 6, 7'"). Another possible strategy is counting on ("There are seven dots. I saw 4 dots and 3 dots, and I said, 'Four, 5, 6, 7'"). In both explanations, the students are using the structure (the organization of the dots) to know when to stop counting. The big difference between the two strategies, however, is in the complexity of the thinking; counting on requires using the idea of set to figure out how many dots there are in all (e.g., four is a set that can be added to; the total number of dots can be thought of as $4 + 3$ or $4 + (1 + 1 + 1)$. This image is accessible to students with a variety of strategies but also encourages students to move beyond a counting-by-ones strategy.

STRUCTURING QUICK IMAGES: USING COLOR TO HIGHLIGHT PART-WHOLE RELATIONS

One way to use more complex quick images like Image B in **Figure 6.4** is to use color to highlight parts of the visual. Intentional use of color supports students in subitizing specific parts of the image, and in doing so, supports them in thinking about the question "How many dots?"

FIGURE 6.5 Structuring quick images: using color

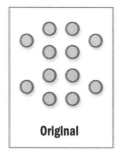 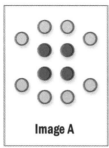 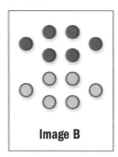 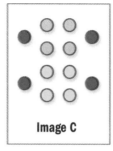 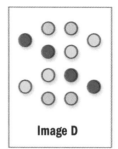

Original   Image A   Image B   Image C   Image D

Take a look at the four versions of the quick image in **Figure 6.5**. Write down how you saw each image. Notice how color influences where you look and how you deconstruct and reconstruct the image. In Image A, the eye is drawn toward the four red dots in the center of the image. In Image B, the eye is drawn to compare two equal sets, which are horizontally divided and symmetrical. In Image C, which is much busier in terms of the use of color, you can visually distinguish three groups of four (or two fours and two twos). Finally, Image D requires more attention to think about the parts. This image might elicit greater variation in students' perception of the image because there many ways to deconstruct it; one person might focus on groups of 2 and someone else might see the diagonal groups of 4 (4 red and 4 yellow) with 2 groups of 2 blue dots, 1 set on the top and 1 set on the bottom. Same image, different use of color, different strategies for answering the question, "How many dots?"

STRUCTURING OUR QUICK IMAGE ROUTINES: THE IMPACT OF QUESTIONS AND PROMPTS

Let's take a look at the image in **Figure 6.6** to help ground our thinking in how the structure of a quick image and the types of questions asked can affect students' mathematical reasoning.

FIGURE 6.6

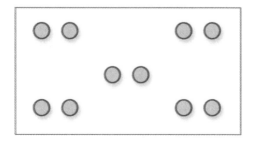

Here are two possible types of question and prompts to use with the image in **Figure 6.6**:

1. "What did you see?" or "Describe what you see" or "Help me see it too!"

2. "How many dots did you see?"

Before reading on, consider this: How might each type of question or prompt produce a different outcome in terms of student reasoning?

Although many of us might think of the second question as the important mathematical question, that question might, at times, actually get in the way of children's mathematical development. Really? How? In our experiences with quick images, asking students, "How many dots did you see?" can nudge them to use a counting-all strategy because they interpret the question as hinting toward a strategy (i.e., "how many" means I should count). Asking, "How many dots did you see?" also tends to make students so focused on the goal (answering the "How many?" question) that they don't slow down to think about how they might use the image's structure to quantify the dots.

Instead, asking, "What did you see?" or saying, "Describe what you see" or "Help me see it too!" can elicit a response that focuses on noticing how the image is structured to describe its organization. Asking students to describe what they see invites them to attend to and use patterns, a critical strategy in problem solving (Schoenfeld 1994). With the image in **Figure 6.6**, students might say, "I saw dots in twos," "I saw 2, 2, 2, 2, 2," "I saw 5 twos," or "I saw 2 twos on the top and 2 twos on the bottom and 2 dots in the middle." Notice how asking students to *describe what they see* puts an emphasis on the structure of the image. When given this kind of prompt, two becomes a fundamental unit that students use to describe the image.

When children mentally structure the image into equal groups, they are developing a big idea in multiplication. In fact, when we take time to have students describe what they see, we have found them more likely to use the grouping structure when we pose paired follow-up questions like "How many dots and how did you figure it out?" For example, children may begin to use early multiplication strategies like doubling (2 and 2 is 4), regrouping the groups (2 and 2 is 4; there's another 2 and 2; so now that's 4 and 4), and skip-counting (2, 4, 6, 8, 10). Although some children may still count the dots by ones (they can often be seen tracing the image with their finger in the air or on the floor in front of them), we have found that there is more variation in students' strategies when we begin with a prompt like "Describe what you see."

Choosing and Designing Quick Images

There are three key questions to keep in mind as you choose and design your own quick images:

1. What's the mathematical goal of my quick image routine?

2. How does the structure of the image support or enhance this goal?

3. How does this goal develop or deepen students' mathematical thinking?

Types of Quick Image Routines

We have found several types of quick image routines to be helpful in developing our students' mathematical reasoning and communication skills. It is important to remember that although we have organized these routines in a certain order within this chapter, this is not the developmental sequence for how you use them. Although we do advocate starting with dot images and asking students to describe what they see (mainly because it slows them down enough to focus on the structure of the image), where you go next with quick images depends on where your students are in their thinking and the kinds of ideas you are hoping to encourage and elicit from them going forward.

Planning Tips

Here are some helpful key questions we always keep in mind when we are planning our routines:

1. What do my students know mathematically and what's my evidence?
   The answers to these questions can usually be found in what students do and say.

2. How will the routine I'm using build upon and challenge students' thinking?

3. What challenges (or misconceptions) might students have and where will they arise?

4. What questions will I use to explore and challenge student thinking and/or misconceptions?

5. How might this particular routine enhance students' ability to communicate their thinking to others?

(For an example of a planning sheet for quick image routines, see **Appendix C**.)

Quick Image Type 1: *Who Is Hiding?*

FIGURE 6.7   In the quick image routine *Who Is Hiding?* the teacher invites students to notice around a slowly revealed picture.

Who Is Hiding? is a quick image routine that is connected to the *Notice and Wonder* routine described in Chapter 1. *Who Is Hiding?* supports students' language development and communication skills. This routine also helps students develop two important problem-solving strategies: (1) slowing down to make sense of the information presented ("What do you notice? Can you describe what you see?"), and (2) using that information to make predictions ("What do you wonder? Who is hiding? How do you know?").

MATERIALS

Although any image can be used in this routine, we suggest that you use images of familiar objects where the parts that make the whole are predictable. For this quick image routine to have the maximum impact, it is important to reveal the image in parts. Unveiling selected parts allows students to predict and revise their predictions based on new information. These quick images can be virtual (images projected on an interactive whiteboard) or physical (cardstock images strategically covered with sticky notes). See the companion website for a slide deck example of the *Who Is Hiding?* quick image routine for use in your classroom.

🎯 GOALS FOR ROUTINE

Who Is Hiding? supports children's descriptive language and ability to use what they notice to make predictions. The goal is not to rush to the final image, but rather to emphasize how interesting it is to have so many different ways to describe the same image. It is also important to highlight to students that what they notice can help them make predictions, and that as they get more information, they can revise their ideas.

ROUTINE STRUCTURE

1. Reveal the image in stages, pausing at each stage for students to notice and make predictions.

2. Begin by having children share their noticings about the image. Ask, "What do you notice?" or "Describe what you see."

3. Record what students say on chart paper or a whiteboard. Because the goal of this routine is to generate as many noticings as possible—for the ideas to *flow*—do not correct or edit students' statements; write students' ideas down as they are said and ask for other noticings.

4. Once children are finished sharing, it is important to highlight the different ways students are *seeing and describing* the same thing ("Wow! Look at all the ways we described this one image—we used words like *brown, fluffy, furry, triangular, ear-like, white and gray, pointing up*").

Quick Image Type 2: *Help Me See It Too!*

Another way to use quick images is for children to develop a repertoire of words that can be used to describe an image precisely. In *Help Me See It Too!* the teacher shows a quick image for a few seconds and students try to describe what they've seen so that everyone can *see this image too*. In this type of quick image routine, it is important for the teacher to "play" with children's noticings in ways that support the development of more and more precise language. (See the Dialogue Box: *Help Me See It Too!* on page 127 for an example of how this routine might be done.)

MATERIALS

Although any image can be used with this type of quick image routine, it is important to remember that the more complex the visual, the more difficult it will be for students to describe accurately and for you to draw. In addition to using basic objects (like dots), we recommend that you keep the objects the same size and color.

GOALS FOR ROUTINE

Help Me See It Too! can be used to support students in developing critical language for describing objects in space. For example, with the image in **Figure 6.8**, students might use words like *left/right*, *top/bottom*, and *corner* to describe what they see (Van Nes and de Lange 2007). Keep track of the words children use (e.g., with a word wall) as a reference tool that will help them develop more precise ways of describing what they see.

FIGURE 6.8 **A word wall of descriptive language used by students in the *Help Me See It Too!* quick image routine**

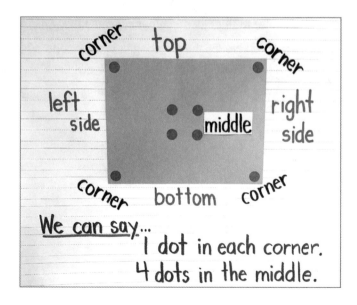

Help Me See It Too!

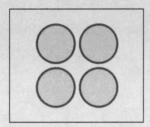

1. **Teacher:** I'm going to show you an image quickly. I'll count to three and when I say three, everyone's eyes should be looking at the screen. Ready? One, two, three. [Shows the image above quickly.] I want you to describe what you saw so that I can see it too.

2. **Naylia:** I saw four blue dots.

3. **Teacher** [draws a rectangle]: So let's make this rectangle the shape of the paper that the image was on. Now I'm going to draw four blue dots [draws them randomly in the rectangle] because that's what Naylia said she saw—four blue dots.

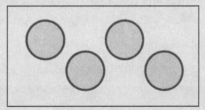

4. **Naylia** [annoyed]: Not like that! It didn't look like that. Let me come up and show you [gets up to go to the board where the teacher is drawing].

5. **Teacher:** Naylia, I'd like you to use your words to help me see it, so please don't come up to draw it for me. I want to be able to picture the image in my mind based on what you say you saw. I heard you say, "four blue dots." [She points to what she's drawn.] That's four blue dots.

6. **Naylia:** There were four blue dots, but it didn't look like that. [She thinks for a moment.] They were in the middle . . . like this [indicates the arrangement of two rows with her hands]. One was on top and one was on the bottom.

(continues)

(continued)

Help Me See It Too!

7. **Teacher:** Oh, the word *middle* is very helpful. [She redraws a rectangle and puts the four blue dots in the middle of the paper.] So four dots in the middle; one was on top and one was on the bottom [draws the following image].

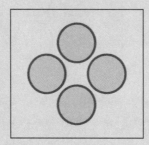

8. **Naylia:** No.

9. **Teacher:** Oh, you said one was on the top and one was on the bottom [points to the blue dot on the top and the one on the bottom].

10. **Naylia:** It looked like a dice—you know how four dots look on dice? [She draws four dots in the air to show where they should be in the image, counting as she indicates the placement with her finger.] This would be one, two, three, four.

11. **Teacher:** I know what four looks like on a die. Like this? [She draws another rectangle making four dots in the shape of a square.]

12. **Naylia:** Yup! That's it.

13. **Teacher:** Hmm. I wonder if there's another way to describe the shape of the dots—if that could help us "see" them the way you saw them.

14. **Naylia:** It looks like a square.

15. **Teacher:** Ah! That's so helpful for me! Now I understand—four dots in the shape of a square! Look at all the words you used to help us [points to each drawing and retells the description]. Here it was four blue dots, but I needed more information to draw the picture exactly. Then you said the dots were in the middle [points to the second drawing], but this dot arrangement wasn't exactly what you saw. When you said the dots looked like four on a die—then I could see what you saw. It was challenging but look at how you persevered and did it! Good for you!

Help Me See It Too!

Several pedagogical moves or behaviors in this short routine are essential to developing students' learning dispositions and communication skills (Mercer 1995). First, the teacher plays with what is said. She pretends that she doesn't know how the image looks and uses the ambiguity in the child's language to challenge Naylia to become clearer in her description. The teacher purposely *creates* a somewhat frustrating situation and, in doing so, helps Naylia communicate more clearly and persist in making herself understood. Second, the teacher does not bounce around from child to child (e.g., saying something like, "Who thinks they can help Naylia be clearer?") but rather stays with one child to support her resilience and communication skills. Third, the teacher uses the board as a tool to support learning; she (1) redraws the image with each new telling; (2) highlights the helpful words Naylia used by writing them down for all to see; and (3) uses what she has drawn to retell Naylia's story. In these ways, the teacher highlights how the language the child is using had become more and more precise.

POSSIBLE EXTENSIONS

A series of images can be used to develop part-whole relations and equivalence (e.g., the different ways to represent a quantity). **Figure 6.9** has a sequence of four images that look different but have the same total number of dots.

FIGURE 6.9   An extension of the *Help Me See It Too!* quick Image routine

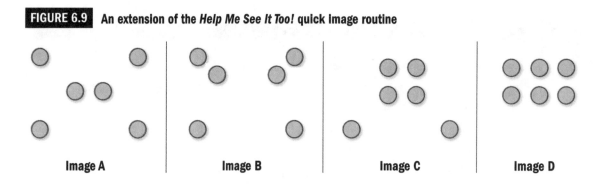

| Image A | Image B | Image C | Image D |

Show each image from **Figure 6.9** separately and have children describe what they see so you can "see" it too. Then, at the end of the routine, put all the images up and ask students, "What do you notice?" As they compare the images, some children may visualize how the dots are changing from one image to another. For example, when comparing Images A and B, a child might say, "The two dots in the middle moved to the top, one on the left and one on the right" (**Figure 6.10**). Other children might recognize that the number of dots stays the same (that all the images have six dots). Some big ideas to highlight with this sequence of images are part-whole relations and equivalence (that 2 + 2 + 1 + 1 = 3 + 3 = 4 + 1 + 1).

FIGURE 6.10

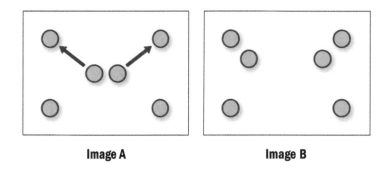

Image A   Image B

Quick Image Type 3: *Comparing Quantities*

Magnitude (an understanding of more and less) is a big idea in early number. Although it is often easy for students to compare sets and say which is more (or who has more), it is more challenging for them to answer the question "How many more cookies does Toni have than Dani?" or to make the two sets of cookies equivalent. The complexity of this question is directly related to some important early childhood math ideas, including cardinality, part-whole relations, and equivalence (Kamii 2000). We have found that using a comparative image (see Chapter 5, *Is It Fair?*) and the questions "How do you know?" and "How can we make it fair?" can support students' reasoning with comparative situations and learning how to create equivalent sets.

MATERIALS

A slide deck of possible *Comparing Quantities* quick images is available on the companion website.

⊙ GOALS FOR ROUTINE

There are three variations of the *Comparing Quantities* quick image routine. Note that although each variation is about comparing sets, the mathematical goals of each variation are different.

VARIATION 1: MORE _____ OR MORE _____? (e.g., "Are there more white dots or more blue dots?")

The goal of this routine is to have children compare two sets. Because the goal is *not* about quantifying how much is in each set, it is important to use images that cannot be easily counted or subitized.

FIGURE 6.11

Image A

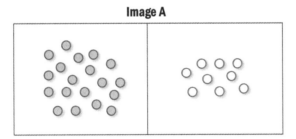

In *More _____ Or More _____?* children reason around which image has more objects.

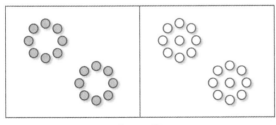

Image B

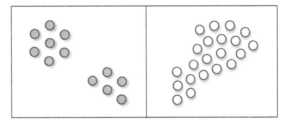

Image C

WHAT'S THE MATH?

Magnitude is an important idea in early number (Sophian 2008). Although it may seem relatively easy for students to visually compare the sets of dots in **Figure 6.11** and answer the question "More blue dots or more white dots?," there are two mathematical big ideas that are embedded in this comparison. First, the sets are not equivalent. Second, to make them equivalent, there needs to be a one-to-one correspondence between the number of blue dots and the number of white dots. As the series of quick images progresses (from Image A to B to C in **Figure 6.11**), it becomes more challenging to know which is more, but there are clues in the organization of the dots to help children (e.g., it's the same image minus one dot in the center).

Although we do not ask children to make the sets equivalent in this routine, it is important to highlight the different ways this picture might be interpreted. Implicit in the statement "There are more blue dots" is its opposite statement—there are fewer white dots! One way to explore implicit logic is to make a number of statements and ask the students if they are true or false. For example, you might say, (1) "There are more blue dots than white dots. True or false?" (2) "There are fewer white dots than blue dots. True or false?" (3) "The number of blue dots and white dots is the same. True or false?" (4) "There are no white dots. True or false?" For examples of this type of quick image, see the quick image slide decks on the companion website.

VARIATION 2: _____ / Not _____ (e.g., three/not three)

The goal of this quick image routine is to have children compare two sets where one is permanently in view (e.g., the three cookies shown at the top of **Figure 6.12**) and the other is shown as a quick image (Images A–D in **Figure 6.12**). Students use thumb signals to indicate whether the set is three or not three (or whatever quantity you have chosen for the routine). Unlike the routine *More* _____ *or More* _____ ?, these images are designed so that students can subitize and compare quantities.

FIGURE 6.12

Possible images for a *Three / Not Three* quick image routine

Original Image

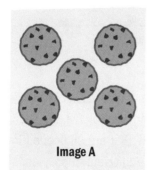

Image A

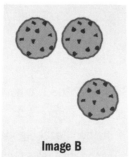

Image B

Image C

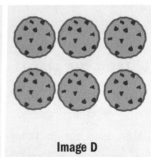

Image D

At the end of the routine, display all the images together and revisit which ones are three and which ones are not three with students. Sort the images accordingly, using a T-chart (**Figure 6.13**) to help students keep track of their comparisons. (Remember, some children may not be convinced that the different arrangements of three are three and may want to count to prove which images are equivalent.)

FIGURE 6.13 After the *Three/Not Three* quick image routine, students sort and compare the images.

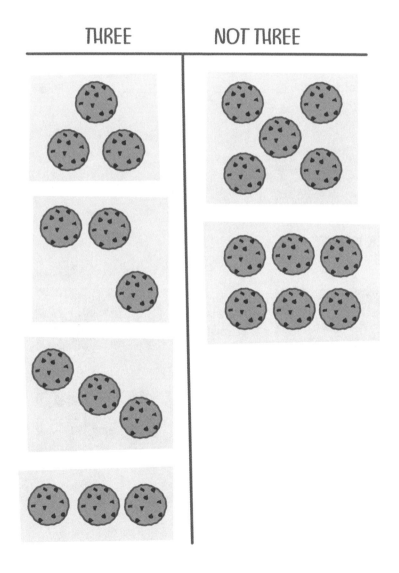

WHAT'S THE MATH?

When students compare Images A–D in **Figure 6.12**, some of them may not be convinced that the different arrangements of three are equivalent sets. For example, some students may describe Image B as being two and one. It may not occur to these children that the two cookies on the bottom and the one cookie on top can actually be put together into a new set called three. The idea that a set can be made up of parts that do not look the same (that the quantity is a constant no matter the arrangement) is a big idea called *conservation* (Kamii 1985). Prekindergarten and kindergarten children often need many opportunities to experience the idea that arrangement does not alter the set or quantity.

EXTENSIONS

Here are a few other ways of using the _____ / *Not* _____ quick image routine:

Ten / Not Ten

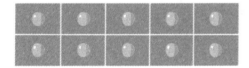

COMPARISON QUICK IMAGES

1.

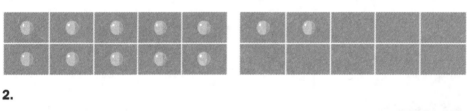

2.

3.

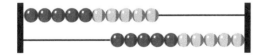

4.

5.

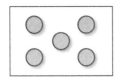 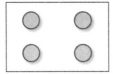

Twenty / Not Twenty

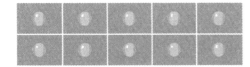 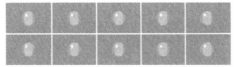

COMPARISON QUICK IMAGES

1.

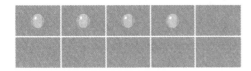 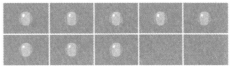

2.

3.

Fifty Blue, Fifty Red

In addition to the 100-frame, we also use the 20-frame as a model. See the companion website for examples of these kinds of lessons.

WHAT'S THE MATH?

This version of the routine using the hundred-frame is particularly useful for first and second graders; however, we have found these images are also appropriate for some younger and older students as well. Although many kindergarten students may still be learning to move beyond counting by ones (to understand cardinality) or are working to learn basic combinations (like early doubles), some kindergarten students *are* ready for the challenge of working with part-whole relations to one hundred.

Why use this routine with the hundred-frame? In our work with children, we have found that exploring the landmark combinations that make one hundred (such as 50 + 50, 60 + 40, 30 + 70) is essential to helping students develop computational fluency in addition and subtraction. In fact, to use a strategy like compensation a child must have a deep working knowledge of landmarks. If a child sees the problem 49 + 51, they might think about this problem in relation to the landmark problem, 50 + 50. Using properties of operations (e.g., the associative and commutative properties), 49 + 51 can be decomposed into 49 + (50 + 1), which is equivalent to 49 + (1 + 50) or (49 + 1) + 50 or 50 + 50. The more we work with the hundred-frame as a visual (using to colors to distinguish the parts that make up the whole of one hundred), the more fluent children become with the landmark facts.

Quick Image Type 4: *How Many ... ?*

One of the most common uses of quick images is to have children answer the question "How many?" Using structured models in the *How Many ... ?* quick image routine supports students in developing strategies for answering this question. These models include the Rekenrek, the bead string, ten-frames, dominoes, and die images. There are many possible ways to use these models as quick images. For more on using the Rekenrek as a quick image, see "Tips for Using the Rekenrek as a

Quick Image" on the companion website. It is important to consider the mathematical ideas you want to develop when you select a model to use for a quick image sequence. For example, here are a few mathematical ideas and strategies you may use quick images to develop:

1. plus one, minus one;

2. counting on;

3. doubles; doubles plus and minus one;

4. part-whole to ten, twenty, or one hundred;

5. unitizing—the ability to shift the unit (grouping) to answer the "How many?" question (e.g., a child sees four filled ten-frames and can answer two questions, "How many dots?" and "How many tens?" The quantity of dots is the same, but if the unit is one, there are forty ones. If the unit is ten, there are four tens).

For quick images that support the development of these mathematical ideas, see the companion website.

Mathematical Models

Children need time to explore the models (e.g., ten-frame, Rekenrek, bead string, etc.) mentioned in the *Comparing Quantities* quick image routines *before* they are used in a quick image routine. If children have been doing the *How Many Days Have We Been in School?* routine (Chapter 2) and an *Attendance* routine (Chapter 3), they will have had many experiences with these models. If they have not done this work, give children the opportunity to explore the models. One way to do this with any model is to show it to children and ask the question "What do you notice?" List students' noticings on the board *without* commenting on what they say. For example, with the Rekenrek, children might use descriptive words like *red, white, beads, balls, row, wood, metal, five red/five white, beads move, two rows, ten on the top, ten on the bottom*. After you create a list of children's noticings, highlight the words that are critical for understanding the structure of the model (e.g., for the Rekenrek, there are five red beads and five white beads on the top row; there are ten beads in each row, etc.). To help students gain additional familiarity with the models, you might also use the practice routines described in Chapter 4.

Assessment: Meeting Student Needs with Quick Images

How do we know what children are ready to learn? All formative assessment starts with developing our ability to notice. What are students doing? How are they doing it? What are they saying? How does what they say and do reflect what they know? Remember, the key here is to find a tool (or tools) that helps you understand what your children know and what ideas they are in the process of developing. Once you have a clear picture of what your children know, you will be able to create routines that build upon and extend their knowledge. In Appendix E we take a look at how one teacher chose quick images to support students' learning.

Using Quick Images to Develop Reasoning Strategies

Quick images are powerful instructional tools that can be used in a variety of ways. In the early childhood classroom, quick images can be used to help young students develop mathematical reasoning and communication as well as counting and addition strategies. And perhaps most importantly, quick images can introduce children to the idea that there are many ways of thinking about a single math problem. Different people will have different ways of thinking about a single quick image, and all of those ways can contribute to the mathematical ideas in our community.

7 Epilogue: Routines as Opportunities to Engage in Playful Learning

" Play is the work of the child. " Maria Montessori, educator

SETTING: Kindergarten Classroom

SCENARIO: The children have been working hard all fall to keep track of the days of school. On this day Toni comes to visit the classroom, and even though it is not the one hundredth day of school, she is wearing a hat to celebrate the one hundredth day (**Figure 7.1**). The children are excited by Toni's hat, but also eager to prove to her how they know it *isn't* the one hundredth day of school! They create thoughtful arguments and use the models that are part of the *How Many Days Have We Been in School?* routine to argue their case. One simple, playful act—lots of very deep learning.

FIGURE 7.1

Toni on the not-one hundredth day of school

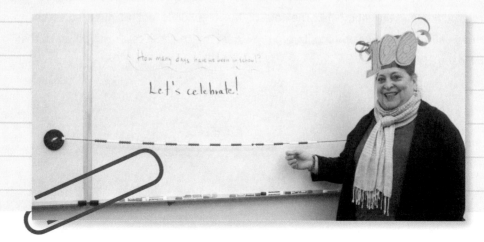

The Power of Play in Engaging Learners in Thinking

Engaging with learners in playful ways changes how they engage with you, as the teacher, and how they experience learning. After Toni's Not-One Hundredth Day visit, one student, Hassan, was so struck by Toni's hat that he pulled his teacher, Emily, aside. Emily wrote to Toni after her visit and said, "Hassan has asked me every few weeks when and if you were going to come back to visit."

In early June, when Toni returned to the classroom, she had an opportunity to work with Hassan again. Emily and Toni created a five-minute one-on-one interview to assess the impact the *How Many Days Have We Been in School?* routine had on students' understanding of place-value big ideas. (See the Chapter 4 material on the companion website to view the assessment we created and to read a transcript of Hassan's conference with Toni.) That morning, before Toni's arrival, Emily hung a hundred bead string in the hallway so that she and Toni would have a quiet space to assess her students' thinking. Emily later told Toni that Hassan, who had noticed the hundred bead string hanging in the hallway prior to the school day starting, approached her and said, excitedly, "Is Ms. Toni here?" When Emily replied, "Yes," he said, "We always play with Ms. Toni. We do math and we play. She makes math so much *fun*. Sometimes she asks me hard questions and it's so much *fun* too. Sometimes Ms. Toni is silly like when she wore the hundredth day hat and it wasn't the hundredth day. That was silly, right?"

What does Hassan's experience tell us? Even though many months had passed since Toni's wearing of the silly one hundredth day hat, Hassan still remembered her visit. What stayed with him about this math experience? It was fun *and* challenging. Hassan's comments are evidence of how memorable playful experiences can be for children. Playful learning creates deep roots in the mind of a learner and can actually change the brain (Brown 2010).

In this book, we advocate for educators to create learning environments designed around intellectual play (Pinchover 2017). We can change *how* we engage with children's minds by changing the types of routines we use and re-envisioning the role of the learner. Because children need ample time to grapple with new ideas, share their thinking, and learn from disagreements, we need to restructure our environments. The role of the teacher is not to push students down artificial learning pathways by a given time, but to give them space to discover the power of their own minds (Donaldson 1986).

Why play? Here are reasons from recent research on how play impacts us (Brown 2010):

- Play is the medium to link mind/brain and body.
- It helps develop and improve social skills.
- It teaches cooperation with others.
- It can heal emotional wounds.
- It can relieve stress.
- It can improve brain function.
- It stimulates mind and body.
- It improves your relationships and your connection to others.
- It keeps you feeling young and energetic.

The routines in this book are one way to create more playful learning environments. To engage children in play, we, as teachers, have to be willing to play ourselves. This may mean examining our curriculum materials and thinking about the types of routines we are doing. The following example is offered as a window into how this might happen.

Using Playful Situations to Assess Students' Fluency

James, a first-grade teacher, had just finished an exploration of doubles and wanted to assess students' fluency. At first, he thought about using timed paper-and-pencil assessment like the one in **Figure 7.2**. However, James recently read an article by Jo Boaler (2014) that outlined why timed tests of basic facts are not effective assessments and can actually be harmful to children. He was at a loss as to how to assess students' understanding of doubles. Timed test? No timed test?

FIGURE 7.2 Doubles Fact Fluency Assessment

2 + 2 =	7 + 7 =	8 + 8 =	10 + 10 =
8 + 7 =	5 + 5 =	3 + 3 =	1 + 1 =
3 + 3 =	8 + 8 =	2 + 3 =	4 + 3 =
2 + 2 =	7 + 7 =	6 + 5 =	6 + 7 =
7 + 7 =	5 + 5 =	2 + 3 =	1 + 1 =
5 + 6 =	6 + 6 =	9 + 9 =	9 + 10 =

Together, Toni and James decided to go a different route and created a routine called *The You-Me Game*. Although the routine would not give him immediate information about his students' *fluency* with doubles, it would give him information about their *understanding* of doubles, which Toni and James decided was more important at this point in the students' learning. Because the game could be used over time and in different ways, it also had the potential to give James much more knowledge about his students' capacity to reason with doubles.

Here is how we introduced *The You-Me Game*:

1. We asked students to sit in a circle.

2. We said, "We are going to play a game today called *The You-Me Game*. The way this game is played is that you will pick a number between one and ten. That's the 'You' number. And once you say your number, I will say my number. That's the 'Me' number."

3. We drew the following chart on the board:

YOU	ME

4. We used the sentence frame "You say _____ ; I [Me] say _____ " after children picked a number. For example, "You say ten; I say twenty." We wrote the numbers on the chart in the appropriate column as we said them.

YOU	ME
10	20

5. After three or four rounds were played, we asked students if they could guess our rule. If they could guess our rule, we asked them *not* to state the rule, but to say a number the "Me" person might say. The goal here was for students to generalize and work backward from the rule they think works (e.g., a child might say, "Write *8* in the 'Me' column; the 'You' number would be four").

We created this game not knowing exactly what to expect from the children. However, we did know we wanted to give them an opportunity to *play* with doubles as ideas. Here's what we learned: within three rounds, many students had figured out our doubles rule! Here's some of the conversation that arose:

Teacher: How many of you like games? [Children shout, "Yay!"] Great! 'Cause I like games too. We're going to play a game today called *The You-Me Game*. Who can think of a number between one and ten?

Student: Five.

Teacher: You say five, I say ten [writes numbers on the "You-Me" chart]. Another number between one and ten?

YOU	ME
5	10

Student: Six.

Teacher: You say six, I say twelve [writes this on the chart]. Another number between one and ten?

YOU	ME
5	10
6	12

Student: Seven.

Teacher: Seven—why are we going in order? So boring [playfully]! [Children laugh.] You say seven, I say fourteen [writes this on the chart].

YOU	ME
5	10
6	12
7	14

Student: I know your rule!

Student: Me too! Me too!

Teacher: Here's a challenge—could you say a number I would write here [indicates the "Me" column]?

Student: Sixteen.

YOU	ME
5	10
6	12
7	14
	16

Teacher: [writes 16 in the "Me" column]: And what will the number in the "You" column be?

Student: Eight.

Teacher: What's the rule?

Student: Eight and eight is sixteen!

Teacher: How would you describe the rule?

Student: The number that you write on the "Me" side would be two times bigger than the one on the "You" side.

Teacher: The number in the "Me" column will be twice as big, or two times bigger than the number in the "You" column. There's another math name for this . . .

Student: That's a double.

Teacher: A double [writes *double* on the board]. It seems like you're ready for another challenge! Okay, here we go. Pick a number between one and ten.

Student: Five.

Teacher: You say five and I say eleven.

YOU	ME
5	11

Student: Seven.

Teacher: You say seven and I say fifteen.

YOU	ME
5	11
7	15

Student: Four.

Teacher: You say four and I say nine.

YOU	ME
5	11
7	15
4	9

Student: I think I know the rule.

Teacher: You know the rule? What could I write here [indicates the "Me" column]?

Student: Thirteen.

YOU	ME
5	11
7	15
4	9
	13

Teacher [writes *13* in the "Me" column]: And if thirteen is the "Me" number, what goes here for the "You" number?

Student: Six.

YOU	ME
5	11
7	15
4	9
6	13

Teacher: What's the rule?

Student: The rule is you double it and then you add one more.

Teacher: So, you double whatever the "You" says and then add one more?

Student [shouting]: Yeah! We got it!!

Teacher [circles the numbers in the "Me" column]: What do all these numbers have in common?

Student: They're all odd.

Teacher: Hmm . . . I wonder why that is—that all the "Me" numbers are odd. Let's keep thinking about that some more tomorrow.

Some of you reading this might be saying, "You still don't know that every child knows their doubles" and you're right. We didn't learn how fluent children were with their doubles combinations. However, what we did learn was more powerful: children's understanding of doubles went way beyond mere recitation of doubles combinations. They were able to use patterns on a chart to generalize a rule for both doubles and doubles plus one and their reasoning was algebraic in nature (Russell et al. 2017). Not only were we stunned by the children, we were amazed at what we didn't know about their thinking. They were so much more capable of playing with numbers than we had imagined! We enjoyed facilitating this routine as teachers, and children loved this game so much that it became a staple of their classroom life. So much so that they even played it with each other during free time!

Our journey in developing the routines we have shared in this book has helped us reimagine our classrooms as *play spaces*. To do this, we have had to examine the kinds of tasks we use with children and ask some fundamental questions:

1. How is this routine going to support and deepen children's thinking?

2. Is there space for children to make mistakes and learn from them?

3. Are there opportunities for students to

   - play with ideas?

   - engage in talk so that they can express their ideas and listen to others' ideas?

   - build connections between what they already know and what they are learning in the routine?

   - create conjectures and generalize big ideas?

4. How will we assess changes in students' thinking and in their learning behaviors?

As we tried to answer these questions, we had to become more creative in designing children's learning experiences. This freedom was both exhilarating and terrifying. We often worried about what children were learning and if it measured up to the expectations set out in our math standards. At the end of the day, we realized that children were learning in deep ways and what they knew actually went well beyond the standard expectations. We also found that children were happier when they experienced math in playful ways, and that when they were happier, we were happier too! We now try to keep one simple but profound axiom in mind: *how children experience learning will forever impact their desire to learn.*

Finally, we would like to share a mantra we use to keep us grounded:

- Be willing to play.

- Be open to learning.

- Be willing to slow down.

- Be willing to *not know.*

- Be willing to make mistakes.

- Be willing to *ask questions* to help you understand.

- Be willing to reflect on your learning.

- Be willing to share your insights.

- Be willing to support and respect the learning of others.

Appendices

Appendix A: A Math Note on *The Calendar*

The Calendar routine is widespread in early childhood classrooms. If you ask educators why, you will get a variety of answers. Rather than enumerating all of the pro and con arguments for using the calendar as a routine, we would like to frame this discussion with two questions: (1) What mathematical ideas are students developing as they work with *The Calendar* routine? and (2) How is the calendar used in the real world?

Let's begin with the second question and think about how the calendar is used in the real world. Picture this: it's Thursday morning and before you leave your house, you're alarmed because you realized you forgot to do the calendar! You remember that you put the date up on Monday, but you were so busy the rest of the week, you forgot to add the appropriate numbers. Happens all the time, right? Maybe it does in school, but in the real world, we don't build the calendar each day. We use the calendar to keep track of events and check the date. You might hear people saying: "Is it the thirteenth or the fourteenth?" or "Did I double book myself?" or "I think I'm pretty booked on Tuesday, but let me check—I might be able to fit you in." If someone pulled out their calendar and asked you, "If today is Monday, what's tomorrow?," you would be totally baffled. This isn't to say that children don't need to know the names of the days of the week and the months of the year. They do. However, this social knowledge can be developed in other, more meaningful ways that do not ask students to sit through a ten- or fifteen-minute *Calendar* routine on a daily basis.

A more challenging question for us to consider is what mathematical ideas we're developing when we use *The Calendar* routine with our students. A common response to this question is "Children learn their numbers." But what exactly does this mean? Are we talking about the sequence of numbers? And is learning the number sequence a big idea or a skill? To answer this important question of what children learn from *The Calendar* routine, we have to think about and compare two kinds of counting: rote and meaningful. Rote counting entails learning the sequence of numbers. With meaningful counting, one learns that numbers can represent quantities and that these quantities have specific relationships to each other. Let's use the numbers fifteen and twenty as examples. Because the calendar is an ordinal model, its numbers do not represent quantities, but a specific day in a month. We don't say October fifteen, we say October fifteenth. With objects, fifteen and twenty can be quantified—they represent a set of fifteen or twenty (this is the big idea of cardinality—the last number counted represents the quantity or set). Each set exists in specific relationships to the other sets. For example, 15 is 5 less than 20 and 20 is 5 more than 15. On the calendar the fifteenth day comes before the twentieth day (or conversely, the twentieth day comes after the fifteenth). However, although the fifteenth day and the twentieth day are *relational*, the relationship is not one of magnitude. The fifteenth day is not *less than* the twentieth and the twentieth day is not *greater than* the fifteenth day. There are twenty-four hours in both!

Children often find the calendar challenging to use as a tool. (Here, we're not talking about saying the names of the days and months.) A tool for what? To answer this question, we have to think about what the calendar *is*, what mathematical ideas underlie its structure. To do so, let's think about four key things: (1) The concept of time is an abstraction and, as such, is enormously challenging for young children to understand (Piaget 1969; Kamii 2012; Russell 2008). (2) In this model, time is cyclic and hierarchical. This means that there are specific units that are embedded in other units (e.g., there are seven days; they form a larger unit called a week; weeks that make up another larger unit called a month; months constitue a year; and so on). (3) The calendar exemplifies a part-whole model where there is a fractional relationship between the days and the number of days in a month. This is straightforward enough until you explore the exact relationship. For example, if today is the fifteenth day of October and there are thirty-one days in October, we *might* think we can represent the relationship between days and number of days in the month with the fraction $\frac{15}{31}$. But here's why that does not work: if it is the fifteenth day, we've completed only fourteen days out of the thirty-one days because we are *still in* the fifteenth day until the twenty-fourth hour has passed! If these ideas are challenging for adults, imagine how difficult these ideas are for young children!

Appendix B: Some Big Mathematical Ideas and Counting Strategies in Early Childhood

Big Mathematical Ideas

These ideas are listed alphabetically and not necessarily in the order children develop them.

- **Cardinality.** The last counted number names the set. Children who have not yet developed this big idea will count a set of objects and then, when asked "How many is that?" will count the set again. The question "How many?" for these children is not about finding a total quantity, but about the act of counting. Another challenge with developing the idea of set is that children often think of the number counted as a label. For example, if there are eight objects in the set and you ask children to show you the eight things, some children will point to the eighth object.

- **Counting.** Here's what has to be coordinated for children to count accurately:
  - *One-to-one tagging*—use some system for isolating what's being counted; it could be touching the person/object being counted or doing this tagging with their eyes (looking at the thing being counted).
  - *Synchrony*—voice is in sync with tagging of students/object counted.
  - *Rote number sequence*—knows the sequence of number names and does not skip or repeat any numbers.

- **Conservation.** The set remains constant even if the objects are moved in such a way that they do not look the same (e.g., objects are moved from a pile into a row).

- **Equivalence.** "Any number, measure, numerical expression, algebraic expression, or equation can be represented in an infinite number of ways that have the same value" (Charles 2005). For example, eight children can arrange themselves in different ways. We could have one group with five students and another group with three students or one group with six students and the other with two students. The total number of students in both situations is equivalent although the number of students in each subset changes.

- **Hierarchical inclusion.** A set is made up of smaller sets. For example, seven is inside eight. This idea is critical for the development of part-whole relations (e.g., 8 = 4 + 4 and 5 + 3).

- **One-to-one correspondence.** This is the underlying big idea that two sets are equivalent (e.g., for twenty-three students we will need twenty-three folders if everyone is going to have one folder).

- **Organizing and keeping track of what's been counted.** This idea includes knowing where you started and where you ended when counting. Often children organize the objects they are counting into groups or rows, or they will move objects from one place to another (e.g., have two piles: things counted and things to be counted).

- **Magnitude.** The comparison of an amount without quantifying (e.g., a child can tell who has more cookies by comparing the quantities: ("He has more!" or "I have less!").

- **Part-whole relations.** Any set can be broken into smaller parts that when combined are equivalent to the whole.

- **Subitizing.** Clements and Sarama (2009) define subitizing in two ways: conceptual and perceptual. *Perceptual subitizing* is the instant visual recognition of small quantities (not more than five things). *Conceptual subitizing* is quickly recognizing smaller groups within a quantity and being able to know the total amount quickly (think of six represented on a die; the two groups of three are easily subitized to make the total quantity, six).

- **Unitizing.** This big idea is one of the critical structures underlying the development of children's multiplicative thinking (Lamon 1996). Unitizing is the ability to simultaneously consider a group of objects as countable in different units. For example, in place value, a child might unitize 100 as 100 ones, 10 tens, or 1 hundred (i.e., $100 = 100 \times 1 = 10 \times 10 = 1 \times 100$).

Strategies for Counting

- **Tagging objects to count.** Children might tag objects, but not have their voice in sync.

- **Moving objects or putting objects in rows (or groups) to count.** Children might put 8 counters into groups of 2 and count, "Two, 4, 6, 8."

- **Counting on.** If there are 8 candies in this tin and you give me 2 more, I don't have to recount the 8 (because it's a set I've already quantified), but I can count on to name the new quantity, [indicating the group of 8 and saying, "Eight, 9, 10"].

Appendix C: Math Routine Planning Sheet

Group size: ☐ small   ☐ whole class

Type of lesson: ☐ game   ☐ mental-math minilesson

Mathematical goal(s): _____

_____

_____

Social goals: _____

_____

_____

Model to be used: _____

_____

Big ideas: _____

_____

_____

Strategies: _____

_____

_____

Possible student struggles: _____

_____

_____

Possible student errors: _____

_____

_____

Teacher Does	Teacher Says	Student Responds	Teacher Writes

Early Childhood Math Routines: Empowering Young Minds to Think, by Antonia Cameron, Patricia Gallahue, and Danielle Iacoviello. Copyright © 2020. Stenhouse Publishers.

Appendix D: A *Notice and Wonder* Routine Around *Drina House in the Morning Mist* with Ms. Casey's First-Grade Class

	Transcript of *Notice and Wonder* Routine	Analysis of Teaching	Implications for Learning
1	**Ms. Casey:** Today, we are going to do something very special.	The teacher begins by emphasizing two words, *notice* and *wonder*. She does this to access her students' prior knowledge. She accepts all answers (even "Wonder Woman") and then clarifies for students what these words mean in the context of the routine they will be doing (lines 1–11).	When teachers accept all student ideas without editing, questioning, or ignoring them, children begin to believe that their ideas matter. Recognizing that one's own ideas give you a kind of power is a critical step in developing autonomous learners, which is foundational in a thinking classroom.
2	**Children** [excitedly]: What?!		
3	**Ms. Casey:** Today, we're going to be learning a new routine called *Notice and Wonder*. Has anyone ever heard the word *notice* before? Haley?		
4	**Haley:** It means to see—to see something.		
5	**Ms. Casey:** To notice is to see. But you know what's so interesting about noticing? We all don't all notice the same things. And that's wonderful. How about the word *wonder*? Has anyone ever heard that word before? And what does it mean?	The teacher highlights that she expects students to notice different things and that divergence (different ideas) are expected, important, and welcome.	Students learn that thinking differently is valued. This sets the norm for a culture where differing ideas can be communicated and heard without judgment.
6	**Sally:** [shouting]: Wonder Woman!		
7	**Ms. Casey:** Okay.		
8	**Sally:** She's wonderful [stands up with her arms in a V] and powerful.		
9	**Ms. Casey:** Thanks for sharing, Sally. That's certainly one way we'd hear the word *wonder*. Wonder Woman. Anyone else?		
10	**Benji:** Wonder is like when you want to know something—you wonder about it.		

	Transcript of *Notice and Wonder* Routine	Analysis of Teaching	Implications for Learning
11	**Ms. Casey:** So, wondering can be a kind of question . . . Okay. So, in our *Notice and Wonder* routine I am going to show you a picture on our screen and I'll show it on the count of three. I'll say, "One, two, three" and on "three," I'll show you the image, the picture. I don't want anyone to say anything aloud right away. I want you to look at the picture and just *notice*. We're going to slow down just to notice. After I show you the picture and give you time to notice, then I'm going to have you just share your noticings. I'm not going to call on people; when you're ready to share your noticing, just say it. We'll popcorn it out.	She lays out the structure of the routine and the expectations for students' learning behaviors in this routine clearly.	When norms are clearly stated for routines, students learn to adjust to the expectations quickly.
12	**Jake** [who has been looking out the window]: Popcorn? Mmm. I love popcorn! [Children giggle.]	The teacher doesn't address the fact that Jake hasn't been paying attention.	Learning how to listen and follow directions is enormously challenging. One way to help children develop listening skills is to challenge them to listen and then paraphrase what they're hearing. These high expectations for students' learning behaviors have huge implications for the development of a classroom community where all voices are heard.
13	**Ms. Casey:** I love popcorn too, Jake. But when I say "popcorn it out," I mean people are going to be sharing their ideas quickly. It's like you say something, then Miles says something, and then Sally. You say it one right after the other—a popcorn-ing out of your ideas. Oh, I just remembered something—please wait until someone's done speaking—we don't talk over each other, but you don't have to raise your hands and wait to be called on. I'll be writing everything you say on our board; your job will be to speak when you're ready—to popcorn out your ideas. Phew! I've just said a lot! Could someone please share what we're doing with our new routine, *Notice and Wonder*? Miles?	She clarifies how the word *popcorn* is used in this metaphor to describe how people will share their ideas—one after the other. She explains what she'll be doing and why she'll be doing it. She gives children think time. She has a highly structured way of doing this routine (e.g., the counting, "All eyes on the picture," the expectation for student silence).	
14	**Miles:** You're going to show us a picture. You're going to say, "One, two, three" and then show us a picture. You don't want us to talk, just notice. When you tell us, we'll share what we notice and you're gonna write what we say on the board.	She uses paraphrasing to restate the task. She highlights the importance of listening and by doing so emphasizes what she values—an important step in developing classroom culture and setting community norms.	

(continues)

(continued)

	Transcript of *Notice and Wonder* Routine	Analysis of Teaching	Implications for Learning
15	**Ms. Casey:** Nice job of listening, Miles! Everyone ready [to the class]? [The children adjust their positions in the meeting area.] Okay; we all look ready. [She says the numbers slowly.] One—all eyes should be looking at our SmartBoard; two . . . three. All eyes on the picture. What do you notice? [Several moments are spent with children looking at the picture.] Okay, when you're ready to share your noticing feel free to speak. [As children speak, Ms. Casey writes what they say as a list on a dry erase board.]	She uses the counting prompt to help children focus on the picture. She records students' noticings in a list on the board, but does not edit their ideas.	Students internalize their role in the routine and learn that the teacher will not accept *hiding out* (i.e., not speaking). Although this is an enormous challenge for some children who are not comfortable speaking in the group, stating the norm and supporting them (the use of wait time) will actually, over time, develop their willingness to contribute. As their confidence grows, so will their willingness to speak.
16	**Child:** I see a house.	Teacher:	
17	**Child:** I see a house.	• notices and names what has happened (only three students have spoken) and sets expectations for learning behaviors—everyone can say something;	
18	**Child:** I see water. [Students grow silent.]		
19	**Ms. Casey:** Hmm. We've only had three people speak and now there's silence. This is something where everyone can speak because everyone can notice something. I'd love to hear everyone's voice. So please relax and share whatever you notice. There are no wrong answers. Everyone's noticing is important.	• addresses the fear ("So please relax") that may be underlying some students' reluctance to speak; • shows that she values everyone's participation and is keeping track of who is speaking.	
20	**Child:** It's a house.	In this segment of the lesson, children are starting to listen to each other and paying attention to what's being written on the board by the teacher. This is an important moment to bookmark as a teacher. When this happens (children noticing repetition or adding details or refining language) it's important to keep track of which children are attending to what's happening. As children get more skilled at listening to each other, there is less repetition of ideas that have been previously said. This	Shifts in students' learning behaviors are one way to assess their development and the impact of a routine. Being able to listen to and understand someone else's ideas is challenging in early childhood. This routine when repeated over time can have a major impact on students' listening and communication skills.
21	**Children:** Someone already said that—it's already up there!		
22	**Child** [responding to her peer]: It's a red house—a tiny red house.		
23	**Child:** The house is in water!		
24	**Child:** The house is on a rock in the water!!		
25	**Child:** I see yellow trees; green ones too.		
26	**Jen:** The people can sit in front—there [points] on that wooden thing.		

(segment continues)

	Transcript of *Notice and Wonder* Routine	Analysis of Teaching	Implications for Learning
27	**Adan** [snickers]**:** That's called a porch.	*(segment continued)* makes the routine go more quickly. It can also nudge children to pay attention to what's written. They begin to keep track of what's been said, compare it to what they're thinking, and, if they're noticing anything new or different, add this to the conversation.	
28	**Ms. Casey:** Let's not correct people. If you'd like to add, please add your idea. You can add something different or build on what someone else has said. Adan, you might have said, "I see a porch." This way you could have helped Jen know that the place in the front of the house has a name, but you could have done it in a way that doesn't feel mean. [Ms. Casey addresses everyone.] So let's remember one of our class agreements [points to their class agreement chart]: *Keep kind in mind.* Let's keep noticing!	The second time the teacher weighs in during the noticing part of the routine is when one child makes fun of another child's lack of knowledge ("That's called a porch"). She uses this occasion to model how the child might have added his idea without being unkind. In doing so, she highlights that she values children's ideas as well as their feelings. To underscore the importance of community, she mentions their agreements as a class ("*Keep kind in mind*"). This is done to remind children that norms have been set for behavior and that these norms are to make everyone feel safe.	Teachable moments can be used to highlight intellectual and social goals for children. For children to internalize these norms, they need to be part of conversations where behaviors are neutrally addressed. Giving children examples of how they might respond in a given situation is a powerful way to correct unwanted behavior. The important thing is that no one—not the child who snickers or the child who didn't know the word *porch*—is publicly shamed.
29	**Child:** There's a porch and people can sit on it.		
30	**Child:** There's a boat.		
31	**Child:** A canoe.		
32	**Child:** Kayak.		
33	**Child:** There's a thing—see [points]—it's round. I don't know what you call it. It's red and white; hanging up on the house. You use it so you don't drown in the water.		
34	**Child:** Oh! I see it now.		

(continues)

159

(continued)

	Transcript of *Notice and Wonder Routine*	Analysis of Teaching	Implications for Learning
35	**Child:** There's stairs in the rock; that must be how they climb up.		
36	**Child:** That's how they got the boat out of the water.		
37	**Child:** There's two windows and a green roof.		
38	**Child:** There's smoke on the water.		
39	**Child:** I think it's called fog—there's fog on part of the water.		
40	**Child:** The house is in the middle of the lake.		
41	**Ms. Casey** [As children have shared their noticings, she has written them down on the board.]**:** Thank you all for sharing your noticings. I'm going to read what I wrote on the board; read with me if you want. [She reads the list of words and phrases.] So many noticings! Now we're just going to think about our *wonderings*. What do you wonder about all the things that we've noticed? [Hands shoot up right away.] Let's give ourselves some time to think. Please put your hands down because we all need time to think about our questions, our wonderings. [Hands go down; she gives students a few minutes to generate questions.] Are we ready? Who would like to start off our wonderings list? Maya?	The teacher uses a talk move, wait time, to ensure that all learners have time to create questions. This simple move provides access to all learners. It also helps children realize that thinking isn't about speed (the person who speaks first isn't viewed as the smartest or the best). Slowing down to create thinking space is a necessary step in developing a culture of ideas.	Understanding takes time. As students realize that they can think about their thinking and refine their ideas and questions by framing them first in their own minds, they begin to develop the kind of self-control necessary for prolonged, deep thought. Think of this habit of mind as a critical tool for developing effective problem-solving strategies.
42	**Maya:** Are we doing the popcorn again?	As children internalize the routine, they are able to name the expectations. Notice how the word *popcorn*, which at first was met with laughter, is now understood as a metaphor for a way of speaking.	

	Transcript of *Notice and Wonder* Routine	Analysis of Teaching	Implications for Learning
43	**Ms. Casey:** Yes. Thanks for reminding us. We'll popcorn out our wonderings. Please wait until I finish writing before you speak, though, because I had a hard time keeping up with everything you noticed.	Notice how the norms are not all set at the beginning as a list, but emerge as the need for them arises. The teacher sets a new norm for *when* to popcorn out—"Please wait until I've finished writing." She also gives students a rationale for what she's asking them to do, which signals that it's important for them to understand her choices— they're not arbitrary.	
44	**Maya:** Why is the house in the water?		
45	**Child:** Who built the house?		
46	**Child:** How could someone build a house in the water?		
47	**Child:** Why would someone want to live there?		
48	**Child:** Do they dive off the porch into the water?		
49	**Child:** Do they fish from the porch?		
50	**Child:** How did they make stairs in a rock? That must have been hard to do.		
51	**Child:** Do they know how to swim or do they have to wear that round thing?		
52	**Child:** Where do they get their food?		
53	**Child:** How many people live there? It's so teeny.		
54	**Child:** Will the house go in the water if there's a big storm?		
55	**Child:** How do they stay warm in the winter?		
56	**Child:** Are they sad by themselves on the lake? Where are their friends?		

(continues)

(continued)

	Transcript of *Notice and Wonder* Routine	Analysis of Teaching	Implications for Learning
57	**Child:** If the water goes up, will it go into the house?		
58	**Child:** Are they scared to be out there by themselves? What if there's lightning?		
59	**Ms. Casey:** Any other wonderings? Okay. This is our first time doing a *Notice and Wonder* routine, but we'll be doing this again. How many of you enjoyed this new routine? [Children give a thumbs up signal.] Here's something we can think about as we continue to use this routine: How do we use what we're noticing to create our questions, to have wonderings? To notice you have to slow down and really observe and think about what you're seeing. Look at all the things we noticed. [She circles the words that describe the house.] Look at all the ways we described the house. It has windows. It's red. It has a porch. It has two windows and a green roof. Now look at some of our questions about the house: "How did someone build a house on the water?" And "Why do they want to live there?" Notice too that we wondered about the safety of the house: Would it fall in the water, and would people be warm? Would they get lonely? What wonderful questions we raised about a picture we might not have even noticed if we hadn't slowed down to pay attention and wonder!	The teacher reframes the purpose of the routine. She revisits the noticings and highlights the words used to describe one thing: a house. She emphasizes complex language structures (e.g., "It has windows. It's red. It has a porch. It has two windows and a green roof.)" She also highlights the importance of slowing down to notice *before* generating questions. The more we notice, the more varied our questions will be. The emphasis on the importance and power of questions is a critical goal for learning.	Over time and with repetition, children's descriptive language improves with the *Notice and Wonder* routine. As they think about what they notice, they begin to use adjectives to give their observations more detail (e.g., "a tiny red house" rather than just "a house"). Notice too that their questions ran the gamut from just three noticings to many. These questions could be explored by children in a small- or whole-group writing exercise. A question like, "Who lives in the house?" could generate a story . . . "Once upon a time, on a house in a lake, lived three children who loved to swim . . ."

Appendix E: Assessment: If This, Then That

How to Use This Chart

As we teach in ways that honor children's development, one of the greatest challenges we have is in knowing when and how to teach what. Here are some key questions:

1. How do I know what a child knows?

2. How can I use what children do or say to guide me?

3. How do I meet children where they are (a "just-right" starting place) and challenge their thinking in ways that shift their understanding?

4. What tools do I have that will guide me?

Our thinking and work in this book and our practice is guided by Fosnot and Dolk's "Landscape of Learning," a model of development in number and operations (2001). Although the "landscape" is not a definitive map, it does give us a rough road map of big mathematical ideas and strategies in number. We can think of this landscape as an "if this, then that" trajectory of children's development. However, we also want to be clear that children's learning is not linear—it's not a dot-to-dot trajectory. Learning is messy; sometimes students make major shifts in their thinking and sometimes they seem to go backward. Because we know that children's thinking can shift radically and in unexpected ways, we want to be prepared to understand how their thinking is changing. The Landscape of Learning is one tool we use to help ourselves assess students' thinking.

We also believe it is important to recognize that there are certain ideas that have to be in place for children to understand other ideas—this is the "if this, then that" schema. For example, if children are learning to count, trying to help them learn to subtract is pointless (and frustrating) because they don't have ideas like set or cardinality in place to understand what makes subtraction work (e.g., a big idea like part-whole relations). This isn't to say that children don't have informal ideas around subtraction from their lives, but rather that formally teaching subtraction as a mathematical structure requires specific big ideas to be in place for children to be able to make sense of it and thrive.

The following chart can help you think about how you might formatively assess your students' mathematical development. Think of this chart as a sketch—it shows one example of how we might assess a group of students' thinking over time, how we would shift the type of quick image we use, and why we would do this. This sketch is one way you could approach the work with these students; there are many possibilities for structuring students' learning, and each of these could be equally valid. Think of our sketch as an architect's drawing—you might question where we put the staircase and, in puzzling over our choices, find an even more effective place for it. We are not giving you the building, but rather the tools with which to draw.

If This, Then That: Assessment Snapshot of Quick Image Routines			
Formative Assessment (What the Teacher Notices, What a Child Says or Does)	**Quick Image Type and Routine Structure**	**What's the Math?**	**What Language Might Be Developed with This Quick Image Routine?**
Assessment 1 During game time, Ms. B. notices that she has a small group of students who always count each pip on a die. They have been playing dice games for several weeks and she hasn't noticed any changes in their counting strategies. Even when she questions them, saying, for example, "Do you need to count the pips on the die, or can you just know it's five?," they still go back to counting when she is not prompting them. To address this group's needs, she decides to use a quick image routine called *Four/Not Four?* She has structured the activity so that all the images that have four look the same (e.g., they look like the pips on a die). She repeats this type of quick image lesson with other die images (e.g., three, six, five, two).	*Comparing images: Four/Not Four?* *Goal of lesson:* Subitizing four as a quantity to move children beyond counting by ones *Structure of the lesson:* • Keep one image up on the board. • Show the other images quickly and ask students to compare the quick images to the stationary one. Students use thumb signals to indicate whether the quantities match (whether there are four dots). • As children give their thumb signals, keep track of any disagreements (e.g., are half the thumbs up and half the thumbs down?). At the end of the lesson, when you reshare the images shown quickly, you might say, "We didn't agree on this image; some of us thought it was four and some of us thought it was not four. I'm going to show it again and let's see if we can agree." • Be sure to encourage talk among the students when there is disagreement. • Sort the images on a T-chart that is labeled *Four* and *Not Four*. After the lesson is over, reiterate what has been learned (or have a student reiterate what they have learned) by saying, "All of these images are four. All of these images are not four. Some are more than four; some are less than four."	*Big ideas* • Cardinality • Equivalence • Magnitude (more/less) • Subitizing • One-to-one correspondence • Conservation	*Possible language* • Same/different • More/less • Four, more than four, less than four • Dots/circles

* This routine type can use any quantity that is on a die. Ms. B. focuses on three, four, five, and six in her series of quick image lessons over the course of several days.

If This, Then That: Assessment Snapshot of Quick Image Routines			
Formative Assessment (What the Teacher Notices, What a Child Says or Does)	**Quick Image Type and Routine Structure**	**What's the Math?**	**What Language Might Be Developed with This Quick Image Routine?**
Assessment 2 Ms. B. notices that her students are now moving away from counting the pips on the die. She wants to challenge their understanding of quantity (or set) by using quick images in which the quantity does not look the same, but may be equivalent. Ms. B. knows that some of her students will grapple with the idea that a quantity can be arranged in different ways and still be equivalent (e.g., they may not be able to recognize that an image that shows three dots in a row and three dots in the shape of a pyramid are the same quantity because they *look* different).	*Comparing images: Four/Not Four?* *Goal of lesson:* Subitizing quantities to compare sets; developing one-to-one correspondence and equivalence *Structure of the images:* • *Do not* put a comparative image up on the board. Rather, say, "I want you all to picture four dots. We are going to be thinking about four dots. I will show you some images and you will show me using your thumb signals whether it's four or not four." • Show the images quickly and ask students, "Four? Not four?" • As children give their thumb signals, keep track of any disagreements (e.g., are half the thumbs up and half down?). At the end of the routine, when you reshare the images, say, "We didn't agree on this image; some of us thought it was four and some of us thought it was not four. I'm going to show it again and let's see what we're thinking now." • Be sure to encourage talk among the students when there is disagreement. • Draw a T-chart and label the two columns *Four* and *Not Four*. • Put all the images that children indicated were four in one column; put the non-four images in the other column. Ask students if they agree or disagree with how the images are sorted. Re-sort the images based on their arguments. • End the routine by highlighting that sometimes four looks like the image on a die [point to that image] and sometimes four *does not* look like the image on the die.	*Big ideas* • Subitizing • Cardinality • One-to-one correspondence • Equivalence • Conservation	*Possible language* • Same/different • More/less • Dots/circles • Row/column • Top/bottom • Right/left

(continues)

(continued)

If This, Then That: Assessment Snapshot of Quick Image Routines			
Formative Assessment (What the Teacher Notices, What a Child Says or Does)	**Quick Image Type and Routine Structure**	**What's the Math?**	**What Language Might Be Developed with This Quick Image Routine?**
Assessment 3 Students have developed fluency with identifying die images and are no longer counting the pips by ones. However, students' articulation of *how* they are seeing the images is often imprecise. Ms. B. decides to work with a quick image routine called *Help Me See It Too!* to help students use more precise language to describe what they see.	***Comparing images: Help Me See It Too!*** *Goal of lesson:* Subitizing quantities and describing the image so someone else can draw it *Structure of the lesson:* • Tell the students you are going to show them an image and their job is to describe it in a way so that you can "see it, too." You will be drawing what they say (how they made you see the image too). • Show the image (e.g., two horizontal dots in each corner of the paper) and ask students to describe what they see so you can see it, too. • Use students' imprecise language to help them learn more precise ways of describing what they see. For example, with two horizontal dots in each corner, expect students to say things like "I saw two dots and two dots and two dots and two dots." Play with the imprecision of their description in your drawing. Initially this may be frustrating for them and they will likely ask to come up and draw it for you. If you keep encouraging students to use their words to describe the image so you can see it, you will find that within one or two sessions their use of descriptive language will improve greatly. If you draw four sets of random groups of two on the board, someone will say, "It didn't look like that; the two dots were in the corner." Highlight how the use of that one word, *corner*, helped you see or know where the dots were. If you draw the two dots in	**Big ideas** • Cardinality • One-to-one correspondence • Equivalence • Subitizing • Conservation • Equal groups • Part-whole relations (e.g., that 8 dots can be thought of as 4 + 4 or 2 + 2 + 2 + 2)	**Possible language** • Dots/circles (or whatever other shapes are used in the image) • Top/bottom • Left/right • Row/column • Horizontal/vertical/ diagonal • Same • Amount • Groups of • Equal groups

(segment continues)

If This, Then That: Assessment Snapshot of Quick Image Routines			
Formative Assessment (What the Teacher Notices, What a Child Says or Does)	**Quick Image Type and Routine Structure**	**What's the Math?**	**What Language Might Be Developed with This Quick Image Routine?**
	(segment continued) one corner vertically, students will say, "It was like this" (showing the horizontal alignment with their hands). You can ask the class to build on each other's ideas by asking if someone knows a word that would describe what the child just indicated with their hand; if no one does, tell them that one math word for what they're showing you is *row*. • Record the words on the board as they come up. At the end of the routine, when the image is drawn, say, "Look at all the words we had to use to describe this image. We said we needed two dots in a row placed in each corner. It was a challenge, but we did it!" • Encourage children's approximations and efforts to be precise in this routine. Although at times *precision* means using a "math term" like *row*, in other cases *precision* can mean using metaphors that accurately describe an image like, "The dots were in the shape of a party hat." The goal here is to honor students' thinking and to develop academic language over time.		

References

Albers, Henry. 2001. *Maria Mitchell: A Life in Journals and Letters.* New York: College Avenue Press.

Anderson, John. 1996. "ACT: A Simple Theory of Complex Cognition." *American Psychologist* 51 (4): 355–368.

Artzt, Alice, Eleanor Armour-Thomas, Frances Curcio, and Theresa Gurl. 2001. *Becoming a Reflective Mathematics Teacher.* New York: Routledge.

Barell, John. 1991. *Teaching for Thoughtfulness: Classroom Strategies to Enhance Intellectual Development.* New York: Longman.

Benoit, Laurent, Henri Lehalle, and Francois Jouen. 2004. "Do Young Children Acquire Number Words Through Subitizing or Counting?" *Cognitive Development* 19: 291–307.

Biddle, Kimberly, Ana Garcia-Nevarez, Wanda Roundtree-Henderson, and Alicia Valero-Kerrick. 2014. *Early Childhood Education: Becoming a Professional.* Thousand Oaks, CA: Sage Publications.

Boaler, Jo. 1993. "The Role of Contexts in the Mathematics Classroom: Do They Make Mathematics More 'Real'?" *For the Learning of Mathematics* 13 (2): 12–17.

———. **2014.** "Research Suggests Timed Tests Cause Math Anxiety." *Teaching Children Mathematics* 20 (8): 469–474.

———. **2015.** *Mathematical Mindsets: Unleashing Students' Potential Through Creative Math, Inspiring Messages and Innovative Teaching.* San Francisco, CA: Jossey-Bass.

Boaler, Jo, and Lang Chen. 2016. "Why Kids Should Use Their Fingers in Math Class." *The Atlantic.* http://www.theatlantic.com/education/archive/2016/04/why-kids-should-use-their-fingers-in-math-class/478053/.

Bormanaki, Himdreza, and Yasin Khoshhal. 2017. "The Role of Equilibration in Piaget's Theory of Cognitive Development and Its Implication for Receptive Skills: A Theoretical Study." *Journal of Language Teaching and Research* 8 (5): 996–1005.

Brabeck, Mary, and Jill Jeffries. 2009. *Effective Practice: Not Drill and Kill.* Online module developed with the APA Coalition for Psychology in Schools and Education. https://www.apadivisions.org/division-7/publications/newsletters/developmental/2019/01/schools-education-coalition.

Bransford, John, Ann Brown, and Rodney Cocking, eds. 2000. *How People Learning: Brain, Mind, Experience, and School.* Washington, DC: National Academy Press.

Brown, Stuart. 2010. *Play: How It Shapes the Brain, Opens the Imagination, and Invigorates the Soul.* New York: Penguin Group.

Brownell, Jeanine, Jie-Qi Chen, Lisa Ginet, Mary Hynes-Berry, Rebecca Itzkowich, Donna Johnson, and Jennifer McCray. 2014. *Big Ideas of Early Mathematics.* Upper Saddle River, NJ: Pearson.

Cameron, Antonia, Diane Jackson, and Betina Zolkower. 1997. *Using the Attendance Routine to Develop Big Mathematical Ideas in Early Childhood.* Unpublished research. New York: Mathematics in the City.

Cash, Richard. 2017. *Advancing Differentiation. Thinking and Learning for the 21st Century.* Minneapolis, MN: Free Spirit.

Chapin, Suzanne, Catherine O'Connor, and Nancy Anderson. 2009. *Classroom Discussions Using Math Talk to Help Students Learn.* Sausalito, CA: Math Solutions.

———. 2013. *Talk Moves: A Teacher's Guide for Using Classroom Discussions in Math.* Sausalito, CA: Math Solutions.

Charles, Randall I. 2005. "Big Ideas and Understandings as the Foundation for Elementary and Middle School Mathematics." *Journal of Mathematics Education Leadership* 7 (3): 9–24.

Clements, Douglas, and Julie Sarama. 2009. *Learning and Teaching Early Math: The Learning Trajectories Approach.* New York: Routledge.

———. 2014. *Learning and Teaching Early Math: The Learning Trajectories Approach.* New York: Routledge.

Cobb, Paul, Terry Wood, and Erna Yackel. 1993. "Discourse, Mathematical Thinking, and Classroom Practice." In *Education and Mind: Institutional, Social, and Developmental Processes,* ed. Norris Minick, Ellice Forman, and Addison Stone. New York: Oxford University Press.

Conklin, Melissa, and Stephanie Sheffield. 2012. *It Makes Sense! Using the Hundreds Chart to Build Number Sense.* Sausalito, CA: Math Solutions.

Costa, Arthur, and Bena Kallick. 2000. *Discovering and Exploring Habits of Mind.* Alexandria, VA: ASCD.

Crowe, Robert, and Jane Kennedy. 2018. *Developing Student Ownership.* Philadelphia: Learning Sciences International.

Csikszentmihalyi, Mihaly. 1991. *Flow.* New York: HarperCollins.

Cuoco, Al, Paul Goldenberg, and June Mark. 1996. "Habits of Mind: An Organizing Principle for Mathematics Curriculum." *Journal of Mathematical Behavior* 14 (4): 375–402.

Danielson, Christopher. 2016. "The Power of Having More Than One Right Answer: Ambiguity in Math Class." *Teaching Children Mathematics* (blog). September 26, 2016. https://www.nctm.org.

Dewey, John. 1944. *Democracy and Education.* New York: Macmillan.

Donaldson, Margaret. 1986. *Children's Minds.* New York: HarperCollins.

Duckworth, Eleanor. 2006. *The Having of Wonderful Ideas.* 3rd ed. New York: Teachers College Press.

Dweck, Carol. 2006. *Mindset: The New Psychology of Success.* New York: Random House.

Edo, Sri, and Zetra Putra. 2011. *Using String Beads to Support Students' Understanding of Positioning Numbers Up to One Hundred.* Conference: Seminar Nasional Pendidikan Universitas Sriwijaya, Palembang, Indonesia, Volume 1. Palembang: Sriwiijaya University.

Emerling, Brad, James Hiebert, and Ron Gallimore. 2015. *Beyond Growth Mindset: Creating Classroom Opportunities for Meaningful Struggle.* Bethesda, MD: Education Week Teacher.

Ergas, Oren. 2017. *Reconstructing "Education" Through Mindful Attention.* London: Macmillan.

Fosnot, Catherine, and Maarten Dolk. 2001. *Young Mathematicians at Work: Constructing Number Sense, Addition and Subtraction.* Portsmouth, NH: Heinemann.

Freudenthal, Hans. 1991. *Revisiting Mathematics Education: China Lectures.* New York: Springer.

Fuson, Karen. 2009. "Avoiding Misinterpretations of Piaget and Vygotsky: Mathematical Teaching Without Learning, Learning Without Teaching or Helpful Learning-Path Teaching?" *Cognitive Development* 24: 343–361.

Gifford, Susan. 2005. *Teaching Mathematics 3–5: Developing Learning in the Foundational Stage.* New York: McGraw-Hill Education.

Ginsburg, Herbert, and Sylvia Opper. 1969. *Piaget's Theory of Intellectual Development: An Introduction.* Upper Saddle River, NJ: Prentice Hall.

Gladwell, Malcolm. 2008. *Outliers.* New York: Little, Brown.

Gonzalez, Norma, Luis Moll, and Cathy Amanti, eds. 2005. *Funds of Knowledge: Theorizing Practices in Households, Communities, and Classrooms.* Mahwah, NJ: Lawrence Erlbaum.

Gravemeijer, Koeno, ed. 1991. "An Instruction-Theoretical Reflection on the Use of Manipulatives." *In Realistic Mathematics Education in Primary School.* Culemborg, The Netherlands: Technipress.

———. 1999. "How Emergent Models May Foster the Constitution of Formal Mathematics." *Mathematical Thinking and Learning* 1 (2): 55–177.

Greenes, Carole, Herbert Ginsburg, and Robert Balfanz. 2014. "Big Math for Little Kids." *Early Childhood Research Quarterly* 19: 159–166.

Groffman, Sidney. 2009. "Subitizing: Vision Therapy for Math Deficits." *Optometry and Vision Development* 40 (4): 229–238.

Guarino, Jody, Rachael Gildea, Christina Cho, and Bethany Lockhart. 2019. "Tools to Support K–2 Students in Mathematical Argumentation." *Teaching Children Mathematics* 25 (4): 208–217.

Hannula, Minna, Pekka Räsänen, and Erno Lehtinen. 2007. "Development of Counting Skills: Role of Spontaneous Focusing on Numerosity and Subitizing-Based Enumeration." *Mathematical Thinking and Learning* 9: 51–57.

Hiebert, James, Thomas Carpenter, Elizabeth Fennema, Karen Fuson, Diana Wearne, Hanlie Murray, Alwyn Olivier, and Piet Human. 2000. *Making Sense: Teaching and Learning Mathematics with Understanding.* Portsmouth, NH: Heinemann.

Jung, Myungwhoon, Paula Hartman, Thomas Smith, and Stephen Wallace, S.R. 2013. "The Effectiveness of Teaching Number Relationships in Preschool." *International Journal of Instruction.* 6 (1): 165–178.

Kamii, Constance. 1985. *Young Children Reinvent Arithmetic: Implications of Piaget's Theory.* New York: Teachers College Press.

———. 1989. *Young Children Continue to Reinvent Arithmetic: 2nd Grade.* New York: Teachers College Press.

———. 2000. *Young Children Reinvent Arithmetic: Implications of Piaget's Theory,* 2nd ed. New York: Teachers College Press.

———. 2012. "Elapsed Time: Why Is It So Difficult to Teach?" *Journal for Research in Mathematics Education* 43 (3): 296–315.

Kang, Sean. 2016. "Spaced Repetition Promotes Efficient and Effective Learning Policy Implications for Instruction." *Policy Insight from the Behavioral and Brain Sciences* 3 (1): 12–19.

Klein, Anton Steven. 1998. *Flexibilization of Mental Arithmetic Strategies on a Different Knowledge Base.* Utrecht, The Netherlands: Freudenthal Institute.

Ladson-Billings, Gloria. 2014. "Culturally Relevant Pedagogy 2.0: a.k.a. the Remix." *Harvard Educational Review* 84 (1): QuestPro, p. 74.

Lamon, Susan. 1996. "The Development of Unitizing: Its Role in Children's Partitioning Strategies." *Journal for Research in Mathematics Education* 27 (2): 170–193.

Lave, Jean C., and Etienne Wenger. 1991. *Situated Learning: Legitimate Peripheral Participation.* New York: Cambridge University Press.

Leinhardt, Gaea, C. Weidman, and K. M. Hammond. 1987. "Introduction and Integration of Classroom Routines by Expert Teachers." *Curriculum Inquiry* 17 (2): 135–175.

Lembke, Erica, and Anne Foegen. 2009. "Identifying Early Numeracy Indicators for Kindergarten and Grade 1 Students." *Learning Disabilities Research & Practice* 24 (1): 12–20.

Mason, John, Leone Burton, and Kaye Stacey. 2010. *Thinking Mathematically.* Essex, England: Pearson Education Limited.

Mead, Margaret. 1928. *Coming of Age in Samoa: A Psychological Study of Primitive Youth for Western Civilization.* New York: HarperCollins.

Mercer, Neil. 1995. *The Guided Construction of Knowledge. Talk Amongst Teachers and Learners.* New York: Multilingual Matters.

National Council of Teachers of Mathematics. 2014. *Principles to Actions: Ensuring Mathematical Success for All.* Reston, VA: NCTM.

Ng, Sharon, and Nirmala Rao. 2010. "Chinese Number Words, Culture, and Mathematics Learning." *Review of Educational Research* 80 (2): 180–206.

Nunes, Terezinha, and Peter Bryant. 1996. *Children Doing Mathematics.* Oxford, England: Blackwell.

———. 2009. *Key Understanding in Mathematics Learning. Paper 2: Understanding Whole Numbers.* London: Nuffield Foundation.

Perkins, David, Shari Tishman, Ron Ritchhart, Kiki Donis, and Al Andrade. 2000. "Intelligence in the Wild: A Dispositional View of Intellectual Traits." *Educational Psychology Review* 12 (3): 269–293.

Piaget, Jean. 1952. *The Child's Conception of Number.* London: Routledge and Kegan Paul.

———. (1946) 1969. *The Child's Conception of Time.* Translated by A. J. Pomerans. London: Routledge and Kegan Paul.

Pinchover, Shulamit. 2017. "The Relation Between Teachers' and Children's Playfulness: A Pilot Study." *Frontiers in Psychology* 8: Article 2214.

Pine, Karen, David Messer, and Kate St. John. 2001. "Children's Misconceptions in Primary Science: A Survey of Teachers' Views." *Research in Science & Technological Education* 19 (1): 79–96.

Proust, Marcel. 1923. "The Captive." *Remembrance of Things Past.* New York: Random House.

Ray, Max. 2013. *Powerful Problem-Solving: Activities Sense Making with the Mathematical Practices.* Portsmouth, NH: Heinemann.

Richardson, Kathy. 2012. *How Children Learn Number Concepts. A Guide to the Critical Learning Phases.* Bellingham, WA: Math Perspectives.

Ritchhart, Ron, and Mark Church. 2011. *Making Thinking Visible: How to Promote Engagement, Understanding, and Independence for All Learners.* San Francisco: Jossey-Bass.

Ritchhart, Ron, Patricia Palmer, Mark Church, and Shari Tishman. 2006. "Thinking Routines: Establishing Patterns of Thinking in the Classroom." Paper presented at AERA. San Francisco, CA, April 7–11, 2006.

Russell, Kelly A. 2008. *Children's Pre-numerical Quantification of Time.* Birmingham: University of Alabama. Unpublished dissertation.

Russell, Susan Jo, Deborah Shifter, Reva Kasman, Virginia Bastable, and Tracy Higgins. 2017. *But Why Does It Work? Mathematical Argument in the Elementary Classroom.* Portsmouth NH: Heinemann.

Salmon, Angela. 2011. "Exploring Young Children's Conceptions About Thinking." *Journal of Research in Early Childhood Education* 25 (4): 364–375.

Sarama, Julie, and Douglas Clements. 2009. *Early Childhood Mathematics Education Research: Learning Trajectories for Young Children.* New York: Routledge.

Schickedanz, Judith, and Catherine Marchant. 2018. *Inside PreK Classrooms,* research by Farran on Tennessee PreK programs. Cambridge, MA: Harvard University Press.

Schoenfeld, Alan. 1994. "Reflections on Doing and Teaching Mathematics." In *Mathematical Thinking and Problem Solving.* Mahwah, NJ: Lawrence Erlbaum.

———. 1998. "Toward a Theory of Teaching-in-Context." *Issues in Education* 4 (1): 1–94.

Schon, Donald. 1983. *The Reflective Practitioner: How Professionals Think in Action.* London: Temple Smith.

Schultz-Ferrell, Karne, Brenda Hammond, and Josepha Robles. 2007. *Introduction to Reasoning and Proof: Grades PreK–2,* ed. Susan O'Connell. The Math Process Standards Series. Portsmouth, NH: Heinemann.

Sherin, Miriam. 2002. "A Balancing Act: Developing a Discourse Community in a Mathematics Classroom." *Journal of Mathematics Teacher Education* 5: 205–233.

Shumway, Jessica. 2011. *Number Sense Routines: Building Numerical Literacy Every Day in Grades K–3.* Portland, ME: Stenhouse.

Simon, Martin. 1995. "Reconstructing Mathematics Pedagogy from a Constructvist Perspective." *Journal for Research in Mathematics Education* 26 (2): 114–145.

Small, Marian. 2010. *Big Ideas from Dr. Small: Creating. A Comfort Zone for Teaching Mathematics, Grades K–3.* Scarborough, Ontario, Canada: Nelson Education.

Smith, Margaret, and Mary Stein. 2011. *Five Practices for Orchestrating Productive Mathematics Discussions.* Reston, VA: NCTM.

Sophian, Catherine. 2008. *The Origins of Mathematical Knowledge in Childhood.* New York: Routledge.

Steffe, Leslie, and Paul Cobb. 1988. *Construction of Arithmetic Meanings and Strategies.* New York: Springer-Verlag.

Stein, Gertrude. 1922. *Geography and Plays.* Boston: Four Seas.

Stephan, Michelle, and Douglas Clements. 2003. "Learning and Teaching Measurement." *Linear and Area Measurement in Prekindergarten to Grade.* NCTM Yearbook. Reston, VA: NCTM.

Steuer, Gabriele, Gisela Rosentritt-Brunn, and Markus Dresel. 2013. "Dealing with Errors in Mathematics Classrooms: Structure and Relevance of Perceived Error Climate." *Contemporary Educational Psychology* 38 (3): 196–210.

Stipek, Deborah. 2018. "Making Circle Time Count." *Development and Research in Early Math Education* (blog). Stanford, CA: DREME.

TERC. 2017. *Investigations in Number, Data, and Space.* 3rd ed. New York: Springer-Verlag. A Complete K–5 Mathematics Curriculum. Glenview, IL: Pearson Education.

Tishman, Shari, David Perkins, and Eileen Jay. 1993. "Teaching Thinking Dispositions: From Transmission to Enculturation." *Theory into Practice* 32 (3): 147–153.

Tomlinson, Carol Ann. 2000. "Differentiation of Instruction in the Elementary Grades." *ERIC Digest.* www.ericdigests.org/2001-2/elementary.html.

———. 2014. *The Differentiated Classroom. Responding to the Needs of All Learners.* Alexandria, VA: ASCD.

Toshalis, Eric, and Michael Nakkula. 2012. "Motivation, Engagement, and Student Voice: The Students at the Center Series." Boston: Jobs for the Future. https://studentsatthecenterhub.org/resource/motivation engagement-and-student-voice/.

Treffers, Adrian. 1991. "Didactical Background of a Mathematics Program for Primary Education." *Realistic Mathematics Education in Primary School,* ed. Leen Streefland. Culemborg, The Netherlands: Technipress.

Valeras, Maria, and Joe Becker. 1997. "Children's Developing Understanding of Place Value: Semiotic Aspects." *Cognition and Instruction* 15 (2): 265–286.

Van den Heuvel-Panhuizen, Marja. 1996. *Assessment and Realistic Mathematics Education.* Culemborg, The Netherlands: Technipress.

———. 2000. *Mathematics Education in the Netherlands: A Guided Tour.* Freudenthal Institute CD-Rom for ICME9. Utrecht, The Netherlands: Utrecht University.

———. 2001. "Realistic Mathematics Education in the Netherlands." *In Principles and Practice in Arithmetic Teaching*, ed. Julia Anghileri. Philadelphia: Open University Press.

———. 2003. "The Didactical Use of Models in Realistic Mathematics Education: An Example from a Longitudinal Trajectory on Percentage." *Educational Studies in Mathematics* 54 (1): 9–35.

———. ed. 2008. *Children Learn Mathematics. A Learning-Teaching Trajectory with Intermediate Attainment Targets for Calculation with Whole Numbers in Primary School.* Rotterdam, The Netherlands: Sense Publishers.

Van Nes, Fenna, and Jan de Lange. 2007. "Mathematics Education and Neurosciences: Relation Spatial Structures to the Development of Spatial Sense and Number Sense." *The Montana Mathematics Enthusiast* 4 (2): 210–229.

Vygotsky, Lev. 1978. "The Role of Play in Development." In *Mind in Society.* Cambridge, MA: Harvard University Press.

West, Lucy. 2014. *Adding Talk to the Equation: A Self Study-Guide for Teachers and Coaches on Improving Math Discussions.* Portland, ME: Stenhouse.

West, Lucy, and Antonia Cameron. 2013. *Agents of Change: How Content Coaching Transforms Teaching and Learning.* Portsmouth, NH: Heinemann.

Wheatley, Grayson. 1999. *Coming to Know Number.* Tallahassee, FL: Mathematics Learning.

———. 2007. *Quick Draw*. Tallahassee, FL: Mathematics Learning.

Index